ANALYTICAL INSTRUMENTATION
PERFORMANCE CHARACTERISTICS AND QUALITY

Analytical Techniques in the Sciences (AnTS)

Series Editor: David J. Ando, Consultant, Dartford, Kent, UK

A series of open learning/distant learning books which covers all of the major analytical techniques and their application in the most important areas of physical, life and materials sciences.

Forthcoming Titles

Polymer Analysis
Barbara H. Stuart, University of Technology, Sydney, Australia

Electroanalytical Chemistry: Principals and Fundamental Concepts
Paul M. S. Monk, Manchester Metropolitan University, Manchester, UK

Environmental Analysis
Roger N. Reeve, University of Sunderland, UK

ANALYTICAL INSTRUMENTATION
PERFORMANCE CHARACTERISTICS AND QUALITY

Graham Currell
University of the West of England at Bristol, Bristol, UK

JOHN WILEY & SONS, LTD
Chichester · New York · Weinheim · Brisbane · Singapore · Toronto

Copyright © 2000 by John Wiley & Sons Ltd,
Baffins Lane, Chichester,
West Sussex PO19 1UD, England

National   01243 779777
International (+44) 1243 779777
e-mail (for orders and customer service enquiries): cs-books@wiley.co.uk
Visit our Home Page on http://www.wiley.co.uk
or http://www.wiley.com

Other Wiley Editorial Offices

John Wiley & Sons, Inc., 605 Third Avenue,
New York, NY 10158-0012, USA

WILEY-VCH Verlag GmbH, Pappelallee 3,
D-69469 Weinheim, Germany

Jacaranda Wiley Ltd, 33 Park Road, Milton,
Queensland 4064, Australia

John Wiley & Sons (Asia) Pte Ltd, 2 Clementi Loop #02-01,
Jin Xing Distripark, Singapore 129809

John Wiley & Sons (Canada) Ltd, 22 Worcester Road,
Rexdale, Ontario M9W 1L1, Canada

British Library Cataloguing in Publication Data

A catalogue record for this book is available from the British Library

ISBN 0 471 99900 8 (HB) 0 471 99901 6 (PB)

To Jenny and Felix for their continued forbearance, and cheerful support

Contents

Series Preface

There has been a rapid expansion in the provision of further education in recent years, which has brought with it the need to provide more flexible methods of teaching in order to satisfy the requirements of an increasingly more diverse type of student. In this respect, the *open learning* approach has proved to be a valuable and effective teaching method, in particular for those students who for a variety of reasons can not pursue full-time traditional courses. As a result, John Wiley & Sons, Ltd first published the Analytical Chemistry by Open Learning (ACOL) series of text books in the late 1980s. This series, which covers all of the major analytical techniques, rapidly established itself as a valuable teaching resource, providing a convenient and flexible means of studying for those people who, on account of their individual circumstances, were not able to take advantage of more conventional methods of education in this particular subject area.

Following upon the success of the ACOL series, which by its very name is predominately concerned with Analytical *Chemistry*, the *Analytical Techniques in the Sciences* (AnTS) series of open learning texts has now been introduced with the aim of providing a broader coverage of the many areas of science in which analytical techniques and methods are now increasingly applied. With this in mind, the AnTS series of texts seeks to provide a range of books which will cover not only the actual techniques themselves, but *also* those scientific disciplines which have a necessary requirement for analytical characterization methods.

Analytical instrumentation continues to increase in sophistication, and as a consequence, the range of materials that can now be almost routinely analyzed has increased accordingly. Books in this series which are concerned with the *techniques* themselves will reflect such advances in analytical instrumentation, while at the same time providing full and detailed discussions of the fundamental concepts and theories of the particular analytical method being considered. Such books will cover a variety of techniques, including general instrumental analysis, spectroscopy, chromatography, electrophoresis, tandem techniques, electroanalytical methods, X-ray analysis and other significant topics. In addition, books in the series will include the *application* of analytical techniques in areas such as environmental science, the life sciences, clinical analysis, food science, forensic analysis, pharmaceutical science, conservation and archaeology, polymer science and general solid-state materials science.

Written by experts in their own particular fields, the books are presented in an easy-to-read, user-friendly style, with each chapter including both learning objectives and summaries of the subject matter being covered. The progress of the reader can be assessed by the use of frequent self-assessment questions (SAQs) and discussion questions (DQs), along with their corresponding reinforcing or remedial responses, which appear regularly throughout the texts. The books are thus eminently suitable both for self-study applications and for forming the basis of industrial company in-house training schemes. Each text also contains a large amount of supplementary material, including bibliographies, lists of acronyms and abbreviations, and tables of SI Units and important physical constants, plus where appropriate, glossaries and references to literature sources.

It is therefore hoped that this present series of text books will prove to be a useful and valuable source of teaching material, both for individual students and for teachers of science courses.

Dave Ando
Dartford, UK

Preface

The central aim of this book is to develop an understanding of the performance characteristics of analytical instruments. Such an understanding is useful in the following:

- the selection of the most appropriate instrument for purchase;
- the confirmation of ongoing performance in order to ensure quality in analytical measurements;
- the estimation of uncertainty in analytical results.

It is possible to develop an understanding of performance characteristics by using either a *bottom-up* understanding of instrumentation and the way in which it works, or a *top-down* appreciation of the needs of analytical science. Both approaches have their own limitations, i.e. the 'bottom-up' approach may fail to link with the real needs of the analyst, while the 'top-down' approach may lack the detail of underlying instrumentation concepts.

In this present text, I have tried to establish a framework that links the analytical *relevance* of performance characteristics across the most common analytical techniques with the physical *limitations* of instrumentation science. With this subject range, it is not possible to concentrate in depth on each topic, and I therefore hope that the compromise that has been struck between breadth and depth will provide a balanced text that is of practical value to the analyst.

A sound understanding of performance characteristics is becoming increasingly important in maintaining *quality* in analytical measurements. Within the 'top-down' approach, I have identified the fact that it is necessary to select an instrument with a level of performance that is appropriate for the given *purpose*, and then it is necessary to implement a quality assurance programme to ensure that the performance *continues* to be satisfactory. In quoting experimental results, it is also essential to be able to assess the performance *uncertainty* in the final results.

There are three main sections to this book. Chapters 1 to 4 establish the context of instrument performance characteristics in relation to analytical methods and the need to establish quality systems in analytical measurement. Chapters 5 to 12 review ('top-down') the performance characteristics appropriate to common analytical systems, while Chapters 13 to 18 address ('bottom-up') specific instrumentation topics that are fundamental to different types of instrument systems.

An understanding of performance by using both of the 'bottom-up' and 'top-down' approaches is an iterative process, which can not be addressed easily in a linear text. Consequently, there are many references in the book that direct the reader to topics appearing in both later, as well as earlier, chapters. In particular, some of the questions require knowledge from other sections and the reader may occasionally have to 'jump ahead' in order to obtain the complete picture.

Chapter 1 establishes the position of the analytical instrument in the analytical measurement process. In Chapter 2, the concept of uncertainty is introduced early in the book in order to emphasize its importance in contributing to a full statement of both instrument performance and analytical results. Performance characteristics themselves are introduced in Chapter 3, which also expands on generic characteristics that are common to many types of instrumentation systems. Chapter 4 provides an overview of the drive for quality systems in analytical measurements.

The book does not seek to detail the design of analytical instruments. Indeed, there are a number of other books (listed in the Bibliography) which describe analytical instruments in very commendable detail. However, Chapters 5 to 12 give brief introductions to the various instrumental techniques as required to support an understanding of the performance characteristics. I have also made use of the various techniques to stress different aspects of the wider subject. The following are some examples of this:

- chromatographic systems which differentiate between the holistic and modular approach to testing instrument performance;

- Fourier-transform infrared (FTIR) spectrophotometry which uses the concepts of the Fourier transform to investigate how spectral information can be encoded in different forms;

- UV–visible spectrophotometry which develops a quantitative evaluation of the variation of measurement uncertainty with the conditions of measurement;

- inductively coupled plasma (ICP)-atomic emission spectroscopy (AES) systems which introduce the use of modern microelectronic detector systems.

Chapter 13 introduces the 'bottom-up' approach by establishing an understanding of *signals*, followed in Chapter 14 by uncertainties that appear in such signals, namely noise, drift and interference. Chapter 15 then introduces the mathematical process of *convolution*, which has applications in understanding signal processes in both hardware and software. The remaining chapters (16–18) are concerned with the characteristics of instrument modules that produce, select and detect electromagnetic radiation.

I believe that the reader will also find it instructive to collect sets of instrument specifications from various manufacturers. Studying these specifications will provide useful practical examples of the use of performance characteristics in specifying the performance of instruments. Some manufacturers also include much useful information about the developments in design and construction that arise from modern advances in technology.

The various questions, both the discussion questions (DQs) and the self-assessment questions (SAQs), are often used to link different instrumental techniques with a common instrumental concept, and it is important that the reader works through all of the questions in order to appreciate the various interrelations that exist. Readers should also use the DQs, as well at the SAQs, to test their understanding of the topics.

I hope that readers will find this book a useful text, and that it will also encourage a greater awareness of performance characteristics, how they are expressed, and how they can be tested.

Graham Currell
University of the West of England, Bristol, UK

Acronyms, Abbreviations and Symbols

AAS	atomic absorption spectroscopy
AC	alternating current
ADC	analogue-to-digital converter
AES	atomic emission spectroscopy
amu	atomic mass unit (dalton)
BP	British Pharmacopoeia
BSI	British Standards Institute
C	coulomb
CCD	charge-coupled device
cCE	chiral capillary electrophoresis
CE	capillary electrophoresis
CEC	capillary electrochromatography
CEN	European Committee for Standardization
CGE	capillary gel electrophoresis
CI	chemical ionization
CID	charge-injection device
CIF	collision-induced fragmentation
CITAC	Co-operation on International Traceability in Analytical Chemistry
CRM	certified reference material
CTD	charge-transfer device
CZE	capillary zone electrophoresis
Da	dalton (atomic mass unit)
DAD	diode-array detector
DC	direct current
DQ	design qualification

DTGS	deuterated triglycine sulfate
DTI	Department of Trade and Industry
ECD	electron-capture detector
EDL	electrodeless discharge lamp
eht	extra-high tension (high voltage)
EI	electron ionization
emf	electromotive force
EOF	electro-osmotic flow
EQ	equipment qualification
eV	electronvolt
F-AAS	flame-atomic absorption spectroscopy
FDA	Food and Drug Administration
FFT	fast Fourier transform
FID	flame-ionization detector
FPD	flame-photometric detector
FTIR	Fourier-transform infrared
FWHH	'full-width-at-half-height'
GC	gas chromatography
GF-AAS	graphite furnace-atomic absorption spectroscopy
GLP	Good Laboratory Practice
HCL	hollow-cathode lamp
HETP	'height of a theoretical plate'
HPLC	high performance liquid chromatography
IC	integrated circuit
ICP	inductively coupled plasma
ID	internal diameter
ILS	instrument lineshape
IQ	installation qualifiication
IQC	internal quality control
IQCAD	internal quality control of analytical data
IR	infrared
ISO	International Organization for Standardization
J	joule
LASER	light amplification by stimulated emission of radiation
LC	liquid chromatography
LED	light-emitting diode
LGC	(formerly) Laboratory of the Government Chemist
LSB	least significant bit
MCT	mercury cadmium telluride
MEKC	micellar electrokinetic capillary electrophoresis
MS	mass spectrometry
MSB	most significant bit
MSD	mass-selective detector

MV	method validation
NAMAS	National Measurement Accreditation Service
NCI	negative chemical ionization
NPD	nitrogen–phosphorus detector
OECD	Organization for Economic Co-operation and Development
opd	optical path difference
OQ	operational qualification
PCI	positive chemical ionization
PDA	photodiode array
PFPD	pulsed-flame-photometric detector
PMT	photomultiplier tube
ppm	parts per million
ppt	parts per thousand
PQ	performance qualification
PT	proficiency testing
PV	performance verification
QC	quality control
RF	radiofrequency
rms	root mean square
sd	standard deviation
SI (units)	Système International (d'Unitès) (International System of Units)
SIM	selected-ion monitoring
S/N	signal-to-noise (ratio)
SOP	Standard Operating Procedure
SST	system suitability testing
TCD	thermal-conductivity detector
TGA	thermal gravimetric (thermogravimetric) analysis
TIC	total ion current
TOF	time-of-flight
UKAS	United Kingdom Accreditation Service
URS	user requirement specification
USP	United States Pharmacopoeia
UV	ultraviolet
V	volt
VAM	Valid Analytical Measurement
VDU	visual display unit
W	watt
a_λ	absorptivity (at wavelength λ)
A	absorbance
b	pathlength
c	speed of light; concentration
CI	confidence interval

CV	coefficient of variation
D	diffusion coefficient
e	electronic charge
E	energy; electric field strength
f	(linear) frequency; focal length
G	responsivity factor
i	current
I	current, light intensity
LOD	limit of detection
LOQ	limit of quantitation
m	mass
m/z	mass-to-charge ratio (mass spectrometry)
MDQ	minimum detectable quantity
P	pressure
Q	electric charge (quantity of electricity)
R	molar gas constant; resistance; resolving power; responsivity
R'	differential responsivity (gain)
RSD	relative standard deviation
s	standard deviation (estimate)
S_A	analytical (output) signal
S_D	output signal offset
S_L	stray light (figure)
S_O	output signal
t	time; Student factor
t_M	void time (in chromatography)
t_m	migration time (in capillary electrophoresis)
T	thermodynamic temperature; transmittance
$u(x)$	uncertainty in x (standard deviation in the measurement)
$\bar{u}$	linear velocity
v_{EOF}	electro-osmotic flow velocity
V	electric potential
$\bar{x}$	mean value of observations
z	ionic charge
δ	optical path difference
λ	wavelength
μ_{eff}	effective (observed) mobility
μ_{EOF}	electro-osmotic flow mobility
μ_0	mean value of distribution
μ_X	mobility of analyte component X
ν	frequency (of radiation)
σ	standard deviation

σ^2 variance

σ_S repeatibility (signal) standard deviation

About the Author

Graham Currell, B.Sc., D.Phil.

After graduating in physics from Southampton University, Graham Currell went on to Oxford University to obtain his doctorate by carrying out research into the bonding of rare earth elements in crystals by using electron parametric resonance spectroscopy. He is now a Principal Lecturer in the Department of Chemical and Physical Sciences at the University of the West of England at Bristol, UK, where his teaching interests have evolved to include applications in scientific instrumentation.

In particular, Graham has for many years been concerned about the problems faced by users of analytical instrumentation in developing countries. Difficulties in maintenance support, the variable electrical and water supplies, and harsh environments are just some of the factors that combine to make the management of sophisticated equipment very difficult in the less-favoured regions of the world. In response to these problems, Graham has developed training programmes in many countries in Asia, Africa, and Central America, for both technical and managerial staff.

The need for a better understanding and quantification of performance characteristics as part of quality assurance systems is now also becoming an increasingly important issue in all countries of the world. Graham is keen to develop training provision in order to enable the users of modern instrumentation to develop a greater understanding of the capabilities, performance and limitations of the equipment that is producing their analytical results.

Chapter 1

Analytical Measurements

Learning Objectives

- To appreciate the quantitative context of instrumental analytical measurements and the different approaches to measurement calibration.
- To differentiate between the role of an analytical instrument and that of a measuring instrument.
- To describe the relationship between the different terms used to describe *uncertainty* and *error*.
- To understand and define the terms used to describe the performance characteristics of an analytical method.

1.1 Analytical Procedures

With such sophisticated instruments is there anything left for the analyst to do?

The aim of an analytical procedure is to discover some specific *information* about the sample under investigation, and ultimately to make a *decision* about that sample.

Such analytical information may be as follows:

- Quantitative, i.e. yielding a *numerical value*, e.g. the concentration of mercury in river water.
- Qualitative, where the analysis identifies a particular *quality* of the sample, e.g. the presence of a particular molecular bonding structure.

In order to obtain this information it is necessary to:

- use an *analytical method* to provide raw results;
- make an *estimate of the uncertainties* in these results;
- (and finally) *interpret* the data.

Ultimately, this information will be used to allow some decision to be made. The need to make a *decision* is an important factor that can not be overemphasized. The instrument may be able to produce raw results, but the analyst must still be able to produce a valid sample, and then be able to interpret the results. This requires a full knowledge of the overall analytical process, including the performance and limitations of the instrument.

1.1.1 Calibration

Most analytical measurements compare the experimental response due to an analyte to that of a *quantitative standard*. The ultimate standards for the basic quantities of mass, length, time, electric current, thermodynamic temperature, amount of substance and luminous intensity are provided by the SI units. However, in routine use, a laboratory will use *derived* standards, e.g. by making up their own solutions of known concentrations or by using certified reference materials (CRMs).

There are two main methods for making measurements by using an instrument, as follows:

- *Direct*, where the output of the instrument is itself calibrated in appropriate units *prior* to the analytical process, e.g. a thermometer or an electronic balance.
- *Comparative*, where the analytical *process* itself carries out the comparison between the analyte response and that of the reference (standard) sample.

For the comparative methods, there are different ways of introducing a standard into the analytical process, for example:

- an *External Standard*, where reference *standards* of known concentration are produced, together with a reference *blank* of zero concentration. The analytical process compares their responses to that of the test sample.
- an *Internal Standard*, where known amounts of the standard are added to the unknown sample, and the *change in response* of the unknown sample is recorded.

The problem in producing a good *external* standard is to achieve a material with similar properties to that of the unknown — concentration, state, and matrix. The effect of the matrix and the extent to which the analyte is bound to the matrix can significantly alter the measurement response.

An *internal* standard may be a material, similar to the analyte, which has a similar but separate experimental response, e.g. a material that gives a separate peak in a chromatogram. The response to the known amount of standard that has been added can be used to calibrate the *response* of the measurement system to the unknown. This process is called *spiking*.

Alternatively, in the method of *standard additions*, the added internal standard is the *same* material as the unknown analyte, and thus increases its measured response. By a process of calculation, it is then possible to deduce how must unknown analyte was originally present.

DQ 1.1

What errors might still occur with the use of an internal standard?

Answer

The advantage of an internal standard is that the matrix is the same as that of the unknown sample. However, there is still some uncertainty as to whether the added material binds with the matrix in the same way as the pre-existing analyte.

1.2 Analytical Instrument

Is an instrument called an 'analytical instrument' just because it is used in an analytical laboratory?

Various types of instrument can be used in the process of analysis. Some are *basic measuring* instruments, such as a thermometer, while others perform an inherent analytical function. In this present book, we will be concentrating on the latter — the *analytical* instruments.

The important distinction between an analytical instrument (e.g. a spectrophotometer) and a basic measuring instrument (e.g. a light-intensity meter) is that:

- an analytical instrument performs an *experiment* as part of the detection process;
- the *conditions* of that experiment can be controlled by the operator to suit the particular analytical situation.

Two examples of analytical instruments that perform an *experiment* are given by (a) a spectrophotometer, which measures the absorbance of radiation passing through a sample, and (b) a chromatograph, which passes the sample through a 'column' within which a chromatographic separation occurs.

The *controls* which are available to the operator in the above two examples are, respectively, as follows:

(a) the wavelength of light used in the spectrophotometer can be systematically changed to provide an output spectrum (absorbance vs. wavelength);

(b) the temperature of the column in a gas chromatograph can be programmed to provide a more suitable chromatogram (detector signal vs. time).

Many *measuring* instruments also perform an 'experiment', e.g. an electronic thermometer may record the emf (voltage) generated by a thermocouple. They may also be very sophisticated and expensive, e.g. electronic balances. However, because the user of the measuring instrument does not have the opportunity of *changing the conditions* of that experiment, they remain basic (albeit *expensive*) measuring instruments.

It is common for analytical instrument systems to contain basic measuring processes as sub-units. For example:

• a spectrophotometer will include a detector unit for the measurement of light intensity;

• the combination of both temperature measurement and mass measurement are essential for the operation of a thermal gravimetric analysis (TGA) system.

DQ 1.2
Is a pH meter, plus electrode, an analytical instrument?

Answer

Certainly an experiment is being performed by using the preferential migration of H^+ ions through the pH glass in the electrode. However, the operator can not alter the conditions of this experiment, and it would be more appropriate, in the context of this present book, to think of the pH meter as a measuring instrument, rather than an analytical instrument.

1.3 Data Output

Is the use of 'chart paper' an essential feature of an analytical instrument?

A simple measuring instrument typically produces a result on a *one-dimensional* (1-D) scale. For example, a thermometer gives a number on a temperature scale, and a balance gives a number on a mass scale. The simple numerical result can then be transcribed directly into a notebook or stored in computer memory.

In an analytical instrument, however, the *conditions* of the experiment can be changed, and these conditions can add *extra dimensions* to the output data. For example, the TGA system gives a *two-dimensional* (2-D) output, with weight plotted against temperature.

The production of multidimensional data is a characteristic of an analytical instrument, and paper has, for a long time, been the accepted medium for the primary output of two-dimensional data in the form of 'charts'. However, with the development of computer technology it is now normal practice to display the 2-D data on a visual display unit (VDU) screen, with the choice of printing 'hard copy' as an option.

The development of modern computer systems has now also enabled the easy visualization of *three-dimensional* (3-D) data by using '3-D' computer displays and printouts. This has encouraged the development of such systems as diode-array detectors for high performance liquid chromatography (HPLC) in which the absorption data are viewed as a function of both time and wavelength.

DQ 1.3

Identify the two (or three) dimensions for the output data obtained from the following analytical instruments. (For example, the UV–visible spectrophotometer gives a 2-D output plot of **absorbance** versus **wavelength**.):

(a) an infrared spectrophotometer;
(b) a flame-atomic absorption spectrometer;
(c) an HPLC system using a refractive-index detector;
(d) a gas chromatography–mass spectrometry (GC–MS) system;
(e) an ICP-AES system.

Answer

(a) transmittance vs. wavenumber;
(b) absorbance integrated over time (at a set wavelength);
(c) changes in refractive index vs. time;
(d) quantity vs. time (GC) and quantity vs. mass (MS);
(e) intensity vs. wavelength.

1.4 Error, Uncertainty and Reliability

No measurement process is perfect, and for each measurement of any variable (e.g. analyte concentration, flow rate and temperature) we can define the following:

True Value. This is the actual value of the variable that we are trying to measure, and is the value that would be obtained if our measurement process was 'perfect'. However, because no process is actually perfect, the 'true value' is *not normally known*. Some scientists are unhappy about the idea of defining a value that can not be known, but for the purposes of this present book it does provide a useful conceptual approach to the problem.

Observed Value. This is the value of the variable given by our measurement process. It is our current estimate of the 'true value'.

Error. The error is the difference between the Observed Value and the True Value, i.e.

Error = Observed Value − True Value

As we do not normally know the 'true value', we do not therefore *know* the size of the actual error in any particular measurement. However, it is important that we are aware that an error is likely to exist and that we can *estimate* how big that error *might* be.

Uncertainty. This is an *estimate* of the possible magnitude of the error. The magnitude of 'uncertainty' must be derived on the basis of knowledge of the way in which the measurement was performed and the reliability of the equipment that was used.

Uncertainty and error are somewhat 'negative-sounding' concepts, and there is considerable concern that they convey unnecessarily negative perceptions to non-scientists. Some analysts prefer to use more positive-sounding terms to describe the same idea, e.g. 'confidence interval', or even 'reliability'.

It is up to the analyst to decide how the *final* result should be presented to the client, and 'confidence interval' may indeed be more acceptable than an expression of 'uncertainty'. However, we shall see in Chapter 2 that, *mathematically*, it is more useful to use the term 'uncertainty' when deriving the estimate of total error because this term is directly related to the standard deviation of the possible outcomes.

The Analytical Methods Committee of the UK Royal Society of Chemistry [1] has discussed the use of the *top-down* and *bottom-up* approaches in estimating analytical uncertainty. A top-down approach aims to use reproducibility data obtained from different laboratories to identify a *de facto* value for the uncertainty, while a bottom-up approach [2] aims to synthesize the total uncertainty by analysing the uncertainties in every step of the process. Both of these approaches have their own strengths and weaknesses.

A top-down approach can not, of itself, identify where the major errors could be occurring in the process. Alternatively, the bottom-up approach may assume unrealistically that certain errors are random and/or independent. It may also miss vital factors, e.g. some forms of laboratory bias (see Section 4.2.1).

DQ 1.4

Which of the following procedures could be used to estimate the error in a new analytical method?

(A) Evaluate the uncertainty by using the performance specifications of the analytical instrument.

(B) Use the new method on a sample whose concentration is already reliably known (CRM).

(C) Spike the sample with a known amount of analyte, and then measure the recovery.

(D) Compare the results obtained by using the new method with those obtained by using an existing method that is already known to have a low uncertainty.

Answer

Method A only addresses errors due to the instrument, and does not include the other uncertainties, e.g. in the sampling procedure. Method B gives the greatest traceability to a standard (true) value of the analyte. However, in cases where CRMs are not available for the analyte or matrix under investigation, it is appropriate to compare (Method D) the new method with an existing method. The use of spiking (C) is a useful procedure that does provide a measure of internal standardization. However, there is still some uncertainty about the extent to which the added analyte binds with the matrix in comparison to the binding of the analyte already present in the sample.

1.4.1 Types of Error

Errors fall into one of the two following major classification types:

Bias (or Systematic Error), where the error (magnitude and sign) remains the same if the measurement is repeated under the same conditions.

Imprecision (or Random Error), where the sign and magnitude of the error change randomly between measurements, even if the measurements are carried out under identical conditions.

These terms all convey the 'negative' concept of uncertainty. Various 'positive' terms are used to give estimates of the maximum errors that are *estimated* to be present in a measurement, as follows:

Trueness, which is an estimation of the **bias** (or systematic error) in the measurement.

Precision (or repeatability), which is an estimation of the **imprecision** (or random error) in the measurement.

Accuracy, which is an estimation of the maximum **total error** in the measurement.

Care should be taken in interpreting the term 'accuracy' as it is often used with different meanings, e.g. it is often equated to precision. It is, however, possible for a measurement with a large systematic error to be very precise, but at the same time to be not at all accurate.

It is easier to detect, and also allow for, the effect of random errors than the effect of systematic errors. A well-designed analytical procedure will, where possible, convert systematic errors into random errors. For example, the process of sampling can introduce serious bias, and care should be taken to develop sampling procedures which randomize the bias.

It is also possible that a particular factor, which gives a systematic error in one laboratory (e.g. an error in a local procedure), may become a random error *between* different laboratories (who use slightly different procedures). This contributes to the fact that the reproducibility of a measurement between laboratories is not as good as the precision of measurements within one laboratory (intermediate precision).

SAQ 1.1

(i) The use of an external reference is required to quantify the following:

    (a) experimental bias   (True/False?);
    (b) experimental imprecision   (True/False?).

(ii) Describe two types of external reference that could be used.

1.4.2 Precision, Repeatability and Reproducibility

The **repeatability** of a measurement is a measure of the precision obtained when that method is repeated under the *same conditions*. This implies the use of the same operator, equipment and reagents, with all of the measurements being made within a short time period.

It is also important to know how the method might vary from day-to-day, but within the *same* laboratory. This is known as **intermediate precision**.

The next stage, namely **reproducibility**, is a measure of the precision of the method when changing *all* of the possible variables, i.e. different operators, reagents, instruments, times and laboratories. The reproducibility is normally acceptable [3] if it is no more than about two to three times greater than the repeatability.

DQ 1.5

An analytical method usually compares the instrumental response of the unknown analyte with that of a standard sample. Which of the following is the most significant **instrument** performance characteristic for this measurement:

(a) accuracy;
(b) trueness;

    (c) repeatability;
    (d) reproducibility;
    (e) intermediate repeatability?

Answer

Note that this question refers to the instrument characteristics and not to those of the method. For the comparative method (Section 1.1.1), it is important that the response of the instrument gives consistent results from run to run, i.e. good repeatability. The accuracy of the **method** *will depend* **primarily** *on the repeatability (precision) of the instrument.*

1.5 Analytical Method Characteristics

We will see in Chapter 4 that, in order to assure quality in analytical measurements, it is necessary to use standard, validated, analytical methods. As part of the validation process, the performance characteristics of the *method* itself must be determined.

The *method* characteristics will include *everything* concerned with the analytical procedure, and is a *top-down* view of the overall performance. Such characteristics record the performance that has been achieved during method validation.

The analytical method will (probably) use an analytical instrument that will have its own *instrument* performance characteristics, where the latter provide *bottom-up* information about *one aspect* of the analytical process. We shall discuss later the relationship between method characteristics and instrument characteristics in the context of different systems (see Sections 5.5 and 8.7).

The characteristics used to *quantify* the performance of an analytical method usually include the following:

- **Accuracy.** Closeness of agreement with the 'true' result. This is the most important parameter for the analytical method (Section 1.4.1).
- **Precision.** Variation in the result when the analytical process is repeated; includes repeatability, intermediate precision, and reproducibility (Sections 1.4.2 and 1.5.1).
- **Range.** The range of concentrations and sample matrices over which the method is valid; this is often determined by the limits of linearity of response. The minimum recommended range [4] depends on the type of analysis, e.g. 80–120% of the expected value for assay.
- **Linearity.** Identifies the range of concentrations over which the method response is *proportional* to the amount of analyte; the linearity of instrument systems is discussed in Section 3.2.4. Note that the straight-line response over the range does not necessarily extrapolate to the origin but would normally be expected to do so.

- **Limit of Detection/Quantitation.** Indication of the minimum analyte concentrations that can be observed/measured; the detectability of instrument systems is discussed in Section 3.3. Clearly, this is an important consideration for trace analysis.
- **Recovery.** Measure of the proportion of analyte detected by the method.

In addition, real samples will often contain other species that may interfere with, or be confused with, the specific analyte being measured. Hence, we must also consider the following:

- **Selectivity/Specificity.** Ability of the method to measure one species of analyte in the presence of other elements or compounds (Section 1.5.3).

A standard analytical method should also be capable of being performed accurately in many different situations, by different operators, using different equipment and reagents from different suppliers, etc. This requirement leads to the concept of the following:

- **Ruggedness/Robustness.** Ability of the method to withstand changes in method, reagents, equipment, operator, etc. (Section 1.5.2).

Finally, when different methods are to be compared, it is important to quantify the overall practical efficiency of the processes, which involves the following aspects:

- **Speed.** The inclusion of the sampling and preparation times, as well as the final laboratory procedure.
- **Cost.** The inclusion of the capital costs of the instruments, as well as their running costs.

Many of these method characteristics reappear in the characteristics of instruments. However, it is important to appreciate that uncertainties in *method* performance do not originate *only* with the instrument, and that other steps in the method will contribute to the total uncertainty.

DQ 1.6

Are 'speed' and 'cost' characteristics that might form part of the 'agreed requirements' between the analyst and client under the UK's initiative on Valid Analytical Measurement (VAM) (see Section 4.8)?

Answer

Both speed and cost are normally very important parameters in any commercial analytical method. There may be 'trade-offs' between accepting a reduced accuracy (which is still sufficient) for the benefit of a faster and cheaper analysis.

1.5.1 Method Precision

A good example of the confusion which exits between the performance character-istics of the *instrument* and those of the *method* is given by the term 'precision'. Although the precision of modern *instruments* can often be very good indeed, the overall precision of the *method* will include many non-instrumental factors such as sampling and extraction where precision can be poor. The overall *method* reproducibility is often considerably worse (see Section 4.2.1) than that of the *instrument* repeatability.

Reviews [5, 6] of many analytical *methods*, over many different analytes, con-centrations, and sample conditions, have shown that, in practice, the expected method reproducibility (the coefficient of variation) (CV) is a function of the analyte concentration, c, and can be given by the following equation:

$$CV\ (\%) = 2^{(1-0.5\log c)} \qquad (1.1)$$

where c is the fractional concentration expressed in powers of ten. This formula gives a good bench-mark for an initial assessment of a new method.

DQ 1.7

A small survey [7] of industrial companies found that typical acceptable errors are quoted as being 0.75% for assay (nominal 95 wt%) and 6.5% for trace (nominal 100 ppm) analyses. Are these in agreement with the prediction of equation (1.1)?

Answer

A concentration of 95% gives a fractional concentration of 0.95 = $10^{-0.022}$, *and equation (1.1) gives* $CV = 2^{(1-0.5(-0.022))} = 2.02\% \approx 2\%$. *A concentration of 100 ppm gives a fractional concentration of* 10^{-4}, *with the equation giving* $CV = 2^{(1-0.5(-4))} = 8\%$. *The quoted errors are therefore within the 'bench-mark' values predicted by equation (1.1).*

SAQ 1.2

Estimate the method precision that could be expected to be obtained for the measurement of an analyte concentration of about 0.5 ppm.

1.5.2 Ruggedness

Ruggedness is a measure of the extent to which small variations in the conditions under which an analytical method is carried out may affect the reproducibility of the method. Tests for ruggedness should seek to quantify the effect of such factors as the following:

- changes in instrumental operating conditions (e.g. pH of the mobile phase);
- use of different instruments (including different columns in HPLC);
- different operators;
- effect of ageing of reagents, etc.;
- different laboratories.

Note that the concept of 'ruggedness' is a function of the *method* and not just the instrument.

DQ 1.8

What is the difference between 'ruggedness' and 'reproducibility'?

Answer

*Testing for ruggedness implies a specific evaluation of the effect of those **separate** aspects of the analytical process that contribute to the **overall** reproducibility figure. This highlights those aspects of the method that are most likely to contribute to possible variations in the result.*

From DQ 1.8, we can see that an evaluation of ruggedness gives an insight into the ability of the method to withstand small changes, and identifies any specific factor variation that may have an exaggerated effect on the result. It is therefore an important contribution to method development.

Ruggedness testing is also useful in that it gives an indication of the instrument response when experimental parameters are reaching their limits. This gives the analyst experience that can be very valuable in trouble-shooting fault conditions in the instrument system.

A full evaluation of 'ruggedness' involves testing the effect of many different experimental variations. There is also the possibility that an interaction between two variations (see Section 3.4) may affect the accuracy of the analysis, although each of these may have little effect independently.

DQ 1.9

Explain why there can be a significant interaction in capillary electrophoresis (Section 11.2) between a decrease in the ionic strength of the buffer and an increase in the applied voltage.

Answer

The link between the two factors is that they both affect the temperature gradient in the capillary. A decrease in ionic strength increases the electro-osmotic flow, thus increasing the current and power dissipation. An increase in voltage also increases the power dissipation. The effect of power dissipation reaches a critical limit (see Section 11.2.2), at which point the temperature would begin to increase dramatically and thus affect the separation performance. On its own, each factor may not reach the limit, although the combined effect may pass it.

1.5.3 Selectivity (and Specificity)

Selectivity and specificity are two terms which are used, almost interchangeably, to describe the extent to which the method can record one parameter exclusively, while being insensitive to a different parameter. A selective (or specific) method does not respond to *unwanted* compounds, and allows only the *wanted* analyte to be measured.

Method selectivity is achieved through good method design and development. It is, however, possible to use the selectivity of the instrument itself (Section 3.2.5). For example, ion-specific electrodes are capable of measuring the presence of particular ions in solution, while remaining insensitive to others.

One distinction, which is used to differentiate between the terms, identifies *selectivity* as the ability to change the *experimental conditions* (e.g. wavelength) to select responses to different analytes, and *specificity* as an *inherent* sensitivity to specific analytes, e.g. a nitrogen–phosphorus detector (NPD) specifically responds to nitrogen and phosphorus.

DQ 1.10

Give an example of **method** selectivity that does not rely on instrument selectivity.

Answer

There are very many chemical procedures (e.g. selective precipitation) that can be used to separate the wanted analyte from unwanted material.

SAQ 1.3

What is the principal characteristic that makes an instrument an *analytical* instrument?

SAQ 1.4

Identify an important characteristic, used to describe analytical methods, which can not be described by a simple *quantitative* parameter.

What is the value in attempting to assess this characteristic?

SAQ 1.5

Explain how *precision* might be the most significant characteristic of an *instrumental measurement*, when the most significant characteristic of the overall *analytical method* is accuracy.

Note. A further question that relates to the topics covered in this chapter can be found in SAQ 5.5.

Summary

This introductory chapter has set the quantitative *context* of analytical measurements, and in so doing has established the 'top-down' theme that runs through the book.

It opened with the fact that analytical measurements are performed so that a *decision* can be made. From this, two main ideas followed, namely that the quantitative process must be calibrated, and that a value must be placed on the possible uncertainty in the result. These ideas were then developed further in the following discussion.

The other main theme was that an analytical method usually requires that an *experiment* be performed, with the analytical instrument enabling the analyst to carry out this experiment under controlled conditions.

Finally, the concept of analytical *method* characteristics was introduced as a way of defining the quality of performance of an analytical procedure.

References

1. Analytical Methods Committee, Analyst, **120**, 2303–2308 (1995).
2. EURACHEM, *Quantifying Uncertainty in Analytical Measurement*. ISBN 0-948926-08-2, Ch. 5, 1995. [Copies available from VAM Helpdesk, LGC (Teddington) Ltd, Teddington, UK.]
3. Maldener, G., *Chromatographia*, **28**, 85–88 (1989).
4. 'Guidance for Industry, QB2 Validation of Analytical Procedures: Methodology', *Proceedings of the International Conference on Harmonization of Technical Requirements for Registration of Pharmaceuticals for Human Use*, ICH-Q2B, Section IV, 1996. [Copies available from Drug Information Branch (HFD-210), Center for Drug Evaluation and Research (CDER), 500 Fishers Lane, Rockville, MD 20857, USA, *or* http://www.fda.gov/cder/guidelines/index.htm.]
5. Horwitz, W., *Anal. Chem.*, **1**, 67A–76A (1982).
6. Boyer, K. W., Horwitz, W. and Albert, R., *Anal. Chem.*, **57**, 454–459 (1985).
7. Groome, S., *VAM Bull.*, **19**, 12–14 (1998).

Chapter 2

Uncertainty and Random Error

Learning Objectives

- To appreciate different statistical descriptions of uncertainty.
- To describe uncertainty in appropriate mathematical formats.
- To calculate the effect of combining uncertainties.

2.1 Introduction

The concept of uncertainty, as being an estimation of possible error, has been introduced in Section 1.4, with errors being categorized as either random (imprecision) or systematic (bias). In this present chapter we look only at the treatment of *random* errors and their estimation through uncertainty.

Where possible, the effect of systematic errors should be randomized and then quantified by making appropriate replicate measurements. For example, the bias present in individual laboratories can be uncovered by collaborative (replicate) trials (see Section 4.2.1) across several laboratories.

The reader is referred to the references at the end of the chapter to provide the background to some of the topics introduced.

2.2 Probability Distribution of Errors

When the wavelength accuracy of a spectrophotometer is quoted as ± 0.5 nm, what is the probability that the error could be 0.6 nm?

Errors can be classified on the basis of their expected probability *distribution* around the true value. The two distributions that are used most commonly are the *normal* distribution and the *rectangular* distribution (Figure 2.1).

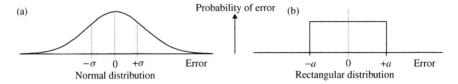

Figure 2.1 Probability of error as a function of error magnitude: (a) normal distribution; (b) rectangular distribution.

2.2.1 Normal Distribution

Random errors in experimental results are usually assumed to follow a normal (or Gaussian) distribution. In this case, a *large* error is less likely to occur than a *small* error (see Figure 2.1(a)). The probability of obtaining a particular value of experimental error is given by the normal distribution, described by the following parameters:

Mean value $= \mu_0$ (error $= 0$)
Standard deviation $= \sigma$
Variance $= \sigma^2$

The mean value, μ_0, is the *true value* of the result, and the 'spread' of possible results is given by the *standard deviation*, σ, of the distribution. The standard deviation is also the root-mean-square (rms) value for the distribution.

If the same measurement were to be repeated (replicated) many times under the same conditions, statistics tells us that we could expect that the following would (approximately) apply:

- 68% of the results would fall within ± 1 standard deviations of the mean;
- 95% of the results would fall within ± 2 standard deviations of the mean;
- 98.8% of the results would fall within ± 2.5 standard deviations of the mean;
- 99.7% of the results would fall within ± 3 standard deviations of the mean.

In practice, for an analytical measurement, the value of μ_0 is *not normally known*, and can only be estimated from the experimental results themselves. The mean of the replicate measurements, $\bar{x}$, is an *estimate* for the value of μ_0.

The true standard deviation, σ, can be *estimated* by either of the following approaches:

- consideration of the detailed process of measurement [1];
- recording the standard deviation, s, of the experimental results.

The parameters *estimated* through the experimental results are as follows:

Mean value of observations $= \bar{x}$
Standard deviation of observations $= s$
Variance of observations $= s^2$

If sufficient repeated measurements can be made, then 's' becomes a very good estimate for the value of σ.

The standard deviation in the observed mean value, '$\bar{x}$', decreases (becomes more accurate) with an increasing number, n, of replicate measurements, and is given by the following equation:

$$u(\bar{x}) = \sigma/\sqrt{n} \qquad (2.1)$$

The confidence interval, CI, estimates the magnitude of the possible error ($\mu_0 - \bar{x}$), in the observed mean, $\bar{x}$, at a chosen level of confidence, via the following equation:

$$CI(p\%) = \pm k_p \sigma/\sqrt{n} \qquad (2.2)$$

where the value of k_p depends on the level of confidence, $p\%$. For example, for a 95% probability that ($\mu_0 - \bar{x}$) falls within the CI, the value, $k_{95} = 1.96$ (≈ 2). If the estimated standard deviation, s, is used instead of σ, then the Student's t-value is used instead of k.

SAQ 2.1

A manufacturer quotes the signal-to-noise (S/N) ratio as being either 10:1 (rms value) or 2:1 (peak-to-peak value). Explain why there is a ratio of '5' between these two values for this parameter.

2.2.2 Rectangular Distribution

There are some situations where the probability of a particular error does not follow a normal distribution [1]. It is possible that all values of error within a specified range are *equally likely* to occur, and that values outside the specified range *do not* occur. The probability distribution of this situation has a rectangular shape (see Figure 2.1(b)). The spread of this type of error is defined by the *limits* of the specified range, e.g. $\pm a$.

An example of where this rectangular distribution of error might arise is in a performance check for the wavelength accuracy of a spectrophotometer. Instruments with errors *anywhere* within an accepted tolerance range would be accepted, but any that fall outside the tolerance would be *rejected* (i.e. a pass/fail condition). This would give the rectangular probability distribution as described.

How can we compare a normal distribution with a rectangular distribution? We will see below in Section 2.4 that combining the effects of different errors as they propagate through the measurement process requires the combination of the *standard deviations* of separate errors. Hence, we need, for the *rectangular* distribution, an *equivalent* 'standard deviation', u_R, that can be used in these calculations. The appropriate relationship derives from the limit values, $\pm a$, and is given by the following equation:

$$u_R = a/\sqrt{3} \qquad (2.3)$$

DQ 2.1

The volume error in a 10 ml pipette is certified as being less than
0.02 ml. Is this uncertainty best described by (a) a normal distribution
or (b) a rectangular distribution?

Answer

*The certification states that the volume will not be outside the specified
limits. As we know nothing about the probability distribution within those
limits, it is appropriate to use the rectangular distribution.*

2.3 Expression of Uncertainty

Why are the ways of stating uncertainty so varied?

The method of quantifying the uncertainty of a particular error will depend on
the probability distribution that the error is expected to follow, i.e.

- For the rectangular distribution, the uncertainty will be given by the limits of
  that distribution, e.g. wavelength accuracy $< \pm 0.5$ nm.
- For the normal distribution, the uncertainty can be expressed as a standard
  deviation (sd), e.g. noise < 0.02 (sd).

The *standard uncertainty*, $u(x)$, in a variable x, is defined as *one* standard deviation in the value of 'x'.

When several uncertainty factors are combined together, the result is a
combined standard uncertainty, $u_C(x)$.

It is often important to give the uncertainty in the value of a variable as a
fraction of the value itself. This fractional uncertainty is frequently expressed as
a percentage, as follows:

$$\text{Coefficient of Variation}(CV(\%)) = \frac{100 \times \text{Standard Deviation}}{\text{Mean}} = 100 \times s/\bar{x},$$

(2.4)

The coefficient of variance is also equal to the Relative Standard Deviation, $RSD(\%)$.

The final result of an analysis should include an estimate of the likely uncertainty range. However, the analyst is still faced with the problem illustrated by
the following discussion question.

DQ 2.2

When quoting an experimental result, should the analyst give the range
of uncertainty (or confidence) which covers:

(a) 68% of possible errors (one standard deviation);
(b) 95% of possible errors, (two standard deviations);
(c) 99.7% of possible errors, (three standard deviations)?

Answer

*The answer depends on the mutual understanding between the analyst and the client, and the criteria used should be clearly stated. In this case, (c) is being over-cautious and would suggest an unnecessarily large error, while (a) gives a high probability (32%) that the true result lies **outside** the quoted error. The usual range covers 95% of probable results, i.e. only a 1 in 20 chance of being wrong.*

The range for the final result should be quoted as an *expanded uncertainty, U,* defined as a multiple of the combined uncertainty, as follows:

$$U = k \times u_C \tag{2.5}$$

where k is the *coverage factor.*

The use of the expanded uncertainty is designed to cover a greater proportion of the possible variation in results. If k is set to '2', then for a true normal distribution, it would cover approximately 95% of the possible variation in results (see DQ 2.2). In other words, with a coverage factor of '2', there is only a 5% (or 1 in 20) chance that the *true* value falls outside the quoted range.

However, where only a few replicate measurements have been made, it would be appropriate to use a higher coverage factor.

SAQ 2.2

The wavelength accuracy for a spectrophotometer is quoted as ± 0.5 nm. Express this as a standard uncertainty.

2.4 Propagation of Errors — Combined Uncertainty

Do random errors just add up, cancel each other out, or behave in another way?

When two or more factors (a, b, c, etc.) give a combined result, t, the uncertainty in the final result, $u(t)$, can be estimated from the individual standard uncertainties ($u(a)$, $u(b)$, $u(c)$, etc.). The method of combining the uncertainties depends on the form of the relationship.

In the derivations given below, we shall assume that there is no interaction between the separate error factors, i.e. each factor contributes independently to the total error.

Applications of the relationships developed in the following sections can be found in the analysis of photometric uncertainty (see Section 5.7).

2.4.1 Addition and Subtraction

For additive relationships of the following form:

$$t = k_a a + k_b b + k_c c \tag{2.6}$$

where k_a, k_b, and k_c are numerical coefficients, it is necessary to use the *absolute* uncertainties. The combined standard uncertainty, $u(t)$, is obtained by taking the squares of the standard deviations of the individual factors (including their coefficients) as follows:

$$u(t) = \sqrt{\{[k_a u(a)]^2 + [k_b u(b)]^2 + [k_c u(c)]^2\}} \tag{2.7}$$

Note that the signs of the 'k' coefficients become irrelevant when each 'k' is squared.

DQ 2.3

In a (poor) gravimetric method, it is necessary to record a small difference between two large mass values. If the uncertainty in each measurement is 0.007 g, estimate the uncertainty in taking the difference between measurements at 2.496 and 2.411 g.

Answer

The calculation is t $= 2.496 - 2.411 \ (= 0.085)$. *This is an expression of the type given in equation (2.6), with* $k_a = +1$ *and* $k_b = -1$. *By using equation (2.7), we get a final uncertainty as follows:*

$$u(t) = \sqrt{[(0.007)^2 + (-0.007)^2]} = 0.01$$

Note that the final result, t $= 0.085$ *(sd of 0.01), has a large fractional error of almost 12%, which is due to the poor practice of calculating a small difference between numbers of similar magnitude.*

SAQ 2.3

The standard deviation in the measurement of an absorbance of value 0.8 in a UV–visible HPLC detector is 0.008, the precision of the autosampler is 0.7% (relative standard deviation) (*RSD*), and the precision of pump flow rate is 0.5% (*RSD*). Estimate the standard uncertainty in the measurement of the peak area.

2.4.2 Multiplication and Division

For relationships of the following form:

$$t = k(a \times b)/c \tag{2.8}$$

where k is a numerical coefficient, it is necessary to use the *fractional* uncertainties. The combined *fractional* standard uncertainty, $u(t)/t$, is obtained by taking the squares of the *fractional* standard deviations of the individual factors, as follows:

$$\frac{u(t)}{t} = \sqrt{\left\{\left[\frac{u(a)}{a}\right]^2 + \left[\frac{u(b)}{b}\right]^2 + \left[\frac{u(c)}{c}\right]^2\right\}} \qquad (2.9)$$

Note that the value of 'k' is irrelevant because of the use of fractional uncertainties. It is possible to use percentage uncertainty ($CV(\%)$) values here, providing that these are used consistently throughout the equation.

2.4.3 Powers

For relationships of the following form:

$$t = ka^n \qquad (2.10)$$

the *fractional* standard uncertainty, $u(t)/t$, is given by the following equation:

$$\frac{u(t)}{t} = n\frac{u(a)}{a} \qquad (2.11)$$

Note that this expression can not be derived by using the multiplicative relationship with repeated 'a' terms (i.e. $a \times a$), because these repeated terms would not be independent.

DQ 2.4

The total radiation power output from a heated filament is given as $E = \sigma AT^4$ (see Section 16.2). If the temperature of the filament increases from 3000 to 3050 K, estimate the percentage increase in radiation output.

Answer

Although this is not directly an 'uncertainty' problem, the mathematics used is the same. The percentage change in temperature is 100 × (3050 − 3000)/3000 = 1.67%. By using equation (2.11), with n = 4, we find that the percentage increase in output = 4 × 1.67 = 6.7%. A small change in the temperature of the source creates a much larger change in the radiation output. This has important consequences for the drift of spectrophotometer light sources (see Section 5.7.2).

2.4.4 Functions

For relationships of a general form where t is a function of a, as follows:

$$t = f(a) \qquad (2.12)$$

then:

$$u(t) = u(a) \times dt/da \qquad (2.13)$$

Note that equation (2.13) can be rewritten in the following way:

Uncertainty in 't' = (Uncertainty in 'a') × (Slope of the variation of 't' with 'a')

We will see a good example of this later in Chapter 3 (DQ 3.9), which estimates the uncertainty in the absorbance, $u(A)$, of an analyte due to a possible uncertainty in the wavelength, $u(\lambda)$, of the measurement, when measured on the *sloping side* of an absorption peak.

DQ 2.5

A calibration curve of absorbance, A, against concentration, c, is given in Figure 2.2. If the standard uncertainty in a measured value of A is 0.05, estimate the standard uncertainty in the equivalent values of the concentration c for measurements of (i) $A = 0.5$, and (ii) $A = 1.0$.

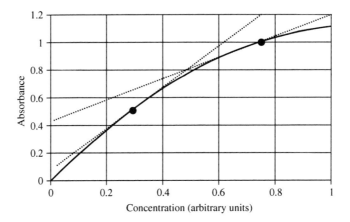

Figure 2.2 Calibration curve of absorbance, A, as a function of concentration, c (DQ 2.5).

Answer

We use here the relationship, $u(c) = u(A) \times dc/dA = u(A)/(dA/dc)$. In this equation, (dA/dc) is the slope of the curve, which can be estimated by drawing tangents to the latter at the relevant values of 'A' (see Figure 2.2), giving slopes for (i) of 1.5 and for (ii) of 0.8; these then give values for $u(c)$ of 0.033 and 0.063, respectively. This question assumes that there is no error in the calibration curve. In practice, however, such errors must be included in the estimation [2].

SAQ 2.4

The uncertainty due to sampling in a particular analysis is 4%, and the instrumental uncertainty is 2%. What overall improvement in the accuracy of the analysis could be obtained if the experimental uncertainty could be reduced to 1%?

SAQ 2.5

The photometric uncertainty of a single-beam spectrophotometer is quoted as $\pm 0.5\%T$, where T is the transmission. Express this uncertainty in terms of a single maximum value of *absorbance uncertainty* within *each* of the two ranges (i) 0 to $0.5A$, and (ii) $0.5A$ to $1.0A$.

Note that $A = -\log(T)$ (see also Section 5.7).

Note. Further questions that relate to the topics covered in this chapter can be found in SAQ 11.3, and DQs 3.9 and 15.6.

Summary

The position of this chapter, i.e. near to the front of the book, has stressed the fact that an appreciation of the uncertainty in any analytical measurement should not be an *afterthought*. The analyst should be quite clear, while performing the analysis, where possible errors might arise and how large these might be.

This chapter has outlined the *quantitative* description of uncertainty, its effect on a final result, and the way in which uncertainties from different sources may be combined.

References

1. EURACHEM, *Quantifying Uncertainty in Analytical Measurement*, ISBN 0-948926-08-2, Ch. 5, 1995. [Copies available from VAM Helpdesk, LGC (Teddington) Ltd, Teddington, UK.]
2. Miller, J. C. and Miller, J. N., *Statistics for Analytical Chemistry*, 3rd Edn, Ellis Horwood, Chichester, UK, 1993, pp. 110–112.

Chapter 3

Instrument Performance Characteristics

Learning Objectives

- To appreciate the reasons for defining instrument performance by a set of quantitative characteristics.
- To identify four different types of instrument performance characteristics.
- To interpret generic instrument response characteristics.
- To differentiate between different types of detectabilily characteristics.
- To appreciate the possibility of interaction between characteristics and the possibility of memory effects in instruments.

3.1 Types of Characteristics

Why are there so many different types of performance characteristics?

The need to specify the performance of an instrument might occur in one of the following situations:

- When we are deciding which instrument to buy.
  Does instrument 'A' make a more accurate measurement than instrument 'B'?
- When we need to check the performance of an instrument.
  Are the results as accurate as they used to be, or has a fault developed?
- When we need to estimate the error of an analysis.
  What is the possible analytical error due to instrumental uncertainty?

A good way to begin learning about performance characteristics is to study the instrument *specifications* provided by the manufacturers. A full set of specifications will include many different performance characteristics, with each one being a way of describing, *quantitatively*, the performance of just one aspect of an instrument (or any other complex system).

SAQ 3.1

What is the difference between an instrument's *specifications* and its *performance characteristics*?

We have seen in Section 1.2 that an analytical instrument *performs an experiment*. It is important to know the performance characteristics of the instrument for the following reasons:

- we can be sure that the instrument is conducting the experiment under *the required conditions*;
- we can *interpret the result* on the basis of knowledge of the response of the instrument;
- we can estimate the *uncertainties* in the result.

A careful examination of a typical set of specifications will reveal that there are FOUR main *types* of performance characteristics for an analytical instrument. These are required to describe how well the experiment is being conducted.

Type I.   The *conditions* under which the instrument can perform the 'experiment'.

Type II.   The *response* of the instrument to the results of that 'experiment'.

Type I(u). The *uncertainties* in the experimental *conditions*.

Type II(u). The *uncertainties* in the *response* of the instrument.

DQ 3.1

Review the following set of specifications for a spectrophotometer, and identify the 'type' of each performance characteristic:

(a) Output display: A and $\%T$;
(b) Range: 350–1000 nm;
(c) Spectral bandwidth: < 5 nm;
(d) Wavelength accuracy: $< \pm 1$ nm;
(e) Photometric accuracy: $< \pm 0.02A$ at $1A$;
(f) Stray light: $< 0.1\%T$ at 220 nm NaI.

Answer

(a) this defines the way in which the instrument responds to the analytical signal — Type II;

(b, c) these define certain experimental conditions — Type I;

(d) an uncertainty in an experimental condition — Type I(u);

(e) an uncertainty in a response — Type II(u);

(f) this gives the maximum uncertainty for stray light (see Section 5.6.3) — although stray light affects the photometric response (Type II(u)), it is also possible to argue that it changes the experimental conditions, i.e. light of a different wavelength is passing through the sample — Type I(u).

3.1.1 Experimental Conditions — Types I and I(u)

These are specific to the type of instrument, for example:

- the spectral bandwidth of a spectrophotometer (Type I);
- the oven-temperature-programme profile for a gas chromatograph (Type I);
- the precision of gradient proportions in HPLC (Type I(u)).

These types of characteristics are discussed in various sections concerning the specific instrumentation system.

3.1.2 Instrument Response — Types II and II(u)

Many response characteristics are *specific* to particular types of instrumentation, and are discussed in specific sections throughout the book. However, there are a number of *generic* response characteristics that reappear in most instruments, with these being discussed below in Section 3.2.

There is also a group of *detectability* characteristics, useful for instruments used in trace analysis, that relate to the minimum analytical signal that the instrument can detect. These are discussed in Section 3.3.

SAQ 3.2

The published specifications of an HPLC pump include the following:

(a) *Flow-rate accuracy* $\pm$ 0.5%;

(b) *Flow precision* $\pm$ 1% (RSD).

Both of the above are given under conditions of *1 ml/min and 1000 psi with water.* Comment on these specifications.

3.2 Generic Response Characteristics

Why do some similar characteristics appear in very different types of instruments?

The following generic characteristics are used to *quantify* the response to analytical signals in many types of instruments:

- *Responsivity* — the ratio of the output signal divided by the input signal (Section 3.2.1);
- *Linearity* — the extent to which the responsivity remains constant for different values of the input (Section 3.2.4);
- *Offset* — the value of the output signal when the input signal is zero (Section 3.2.3);
- *Drift* — the gradual random change of the output signal with time when the input remains constant (Section 3.2.3);
- *Noise* — the rapid random change of the output signal with time when the input remains constant (Section 3.2.2);
- *Dynamic Range* — the ratio of the largest signal that can be measured to the smallest signal (Section 3.2.4);
- *Selectivity* — the differential response to a wanted analyte compared to that of an unwanted analyte (Section 3.2.5).

These terms define the response of the instrument with respect to the following:

- different *levels* of analytical signal (responsivity, linearity, offset and selectivity);
- variations with *time* (drift and noise);
- the *range* over which the instrument will operate (linearity and dynamic range).

The characteristics can be understood by thinking of the instrument as a 'signal-processing' unit — see Figure 3.1. The unknown analytical signal, S_A, is applied to the input of the unit, which then responds by producing an output signal, S_O.

$$S_A \longrightarrow \boxed{\begin{array}{c} \text{Signal} \\ \text{processor} \end{array}} \longrightarrow S_O$$

Figure 3.1 Schematic representation of the use of an instrument as a signal-processing unit.

The relationship, $S_O = f(S_A)$, is known as the 'response function' or 'transfer function'. The signals, S_O and S_A, occur in many *different* forms. For example, S_O may be the output *voltage* from a pH electrode corresponding to the *pH* of the solution, S_A, as given in the following equation:

$$E = E_0 - \frac{2.3RT}{F}(\text{pH} - 7.0) \qquad (3.1)$$

where the output signal $(S_O) = E$ (emf generated by the pH electrode), the input signal $(S_A) = \text{pH}$ (pH of the solution), R is the molar gas constant, F is the Faraday constant and T is the temperature (K).

In the case of an electronic amplifier, the two signals are often of the same form, i.e. voltages. It is also possible that the output of an instrument is *calibrated* in the same units as the variable that it is measuring, e.g. some spectrophotometers can be calibrated to give an output reading directly in concentration units. The *response function* (or *transfer function*) can be expressed by using the following:

(i) An *algebraic equation*, e.g. equation (3.1) which gives the emf obtained from a pH electrode as a function of the pH of the solution.

(ii) A *table of corresponding values* of S_O and S_A. This method is used in computers as a 'look-up' table to convert quickly from one variable to another. However, this can be a slow process when performed *without* the use of computers.

(iii) A *graph* of S_O versus S_A, e.g. Figure 3.2. This has the advantage of giving a quick *visual appreciation* of the response.

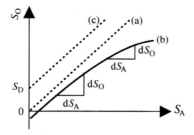

Figure 3.2 Illustration of an instrument response function.

The ideal transfer function would normally be the linear response, which goes through the origin of the graph (zero offset), as shown by line (a) in Figure 3.2. Line (b) shows an instrument response that is non-linear (it is not a straight line) and which has a non-zero offset (i.e. the line does not pass through the origin). The offset (S_D) is normally measured as the magnitude of the intercept on the ordinate (y) axis. In the case of line (b) the offset is actually negative.

3.2.1 Responsivity

The common mathematical form of a straight line (linear response) is given by the following equation:

$$y = mx + c \qquad (3.2)$$

It is possible to define the straight line by using only the following two variables:

- the slope, m, which is given by $m = dy/dx$;
- the intercept, c, which is the offset on the y-axis.

Given these two values we know exactly which line must be drawn.

The equivalent equation for the straight-line response of an instrument with a non-zero offset (line (c) in Figure 3.2) is as follows:

$$S_O = R'S_A + S_D \qquad (3.3)$$

For the ideal instrument response, the offset, S_D, should be zero.

The *slope* of the line is the *differential responsivity* (or *gain*), R', which is given by the following:

$$\text{Differential Responsivity } (R') = \frac{\text{Change in } S_O}{\text{Change in } S_A} = \frac{dS_O}{dS_A} \qquad (3.4)$$

The differential responsivity is defined at *specific points* on the graph.

However, *responsivity* is defined simply as follows:

$$\text{Responsivity } (R) = \frac{\text{Ouput Signal}(S_O)}{\text{Input Signal}(S_A)} = \frac{S_O}{S_A} \qquad (3.5)$$

Responsivity is usually the key parameter for an instrument system. It measures the magnitude of the observed output value, S_O, as a function of the input analytical signal, S_A. It is important not to confuse responsivity, $R = S_O/S_A$, with *differential* responsivity (or slope). For a linear response (i.e. a straight line with constant slope), the differential responsivity will be the same for all values of S_A. It will also be *independent of any offset*. For example, both line (a) and line (c) in Figure 3.2 have the same slope, and hence the same value of R', at all values of S_A.

The responsivity (R), however, is the direct ratio of S_O and S_A (not their changes) and will therefore be affected by any offset value added to S_O. Hence R' and R are *only the same* if the *offset is zero*. We can therefore write the following relationship:

$$R = \frac{S_O}{S_A} = \frac{R'S_A + S_D}{S_A} = R' + \frac{S_D}{S_A} \qquad (3.6)$$

For a *non-linear* graph, the differential responsivity (or slope) of the graph will also vary from point to point on the graph — see line (b) in Figure 3.2.

The term 'sensitivity' is sometimes regarded as being equivalent to responsivity. However, 'sensitivity' is also often used as a measure of the minimum detectable signal, and this can lead to confusion in its use — see Section 3.3.

DQ 3.2

Calculate the differential responsivity of the pH electrode at 25°C.

Answer

*Compare the equations (3.1), (3.2) and (3.3). The **slope** of the line, E vs. pH, is given by -2.3 RT/F. By substituting the appropriate values* *(remembering that temperature must be in degrees absolute), we find that the differential responsivity (slope) of the pH electrode at 25°C = −59 mV pH⁻¹.*

3.2.2 Noise

Noise is a fundamental problem in any form of measurement and appears in a variety of forms, i.e. drift, interference, and both short-term and long-term noise.

The identification of the type of noise, and the methods available for reducing it, are treated as a specific topic in Chapter 14.

3.2.3 Offset and Drift

Offset (or zero error), S_D, occurs if a zero-input analytical signal results in a non-zero output signal — see lines (b) and (c) in Figure 3.2. Offset may occur due to the following:

- an interfering signal from some other component of the sample;
- a standing signal in the instrument system.

Where possible, instruments are designed so that any standing signal, arising in the instrument system, can be offset to zero electronically, by adding or subtracting an equivalent signal in the output. For example, the control used to set 0%T (or ∞A) in some simple spectrophotometers ensures that the output *reads* 0%T when no light is falling on the detector.

Some systems have an inherent offset in the experimental process. Again, the pH electrode (equation (3.1)) provides a good example in that there is an offset, S_D, when pH = 0, which is given by the following:

$$S_D = E_0 - (2.3RT/F)(-7.0) = E_0 + 16.1RT/F \qquad (3.7)$$

DQ 3.3

The responses (differential responsivity and offset) of a pH meter need to be adjusted so that its performance matches the theoretical equation. Most pH meters provide controls to adjust these parameters — identify which controls adjust which parameters in the following table:

Control	Differential responsivity	Offset
'Set temperature'		
'Set buffer'		

Answer

We can see from DQ 3.2 that the temperature affects the differential responsivity of the electrode. Hence, the 'set-temperature' control adjusts *the differential responsivity (slope) of the meter to compensate for the change in performance of the electrode due to different temperatures. In addition, some pH meters have a separate 'Slope' control, which also adjusts the differential responsivity (slope) of the instrument to compensate for changes in the differential responsivity of the electrode with age. The 'set-buffer' control shifts the offset, S_D, without changing the differential responsivity.*

By using these two controls, the straight-line response can be adjusted to the correct calibrated position in the S_A/S_O graph.

The offset setting in instrument systems is particularly susceptible to *drift* (see Section 14.1). Examples where drift can affect the offset include the following:

- a change in the intensity of a spectrophotometer source leads to a *drift* in the 0A setting, unless this is compensated by recalibration against the Reference Sample;
- a drift in the offset of a pH measurement requires the resetting of the 'Set-Buffer' control;
- a drift in the '0-volt' output from a DC amplifier requires a readjustment of the 'Set-Zero' control.

Drift, leading to a non-zero offset, is a low frequency example of '$1/f$' noise (Section 14.3).

3.2.4 Linearity and Linear Dynamic Range

Where possible, an analytical measurement is designed to give a linear response with zero offset, as shown by line (a) in Figure 3.3. However, no system is perfect, and as an example, the line (b) in Figure 3.3 has been drawn from data

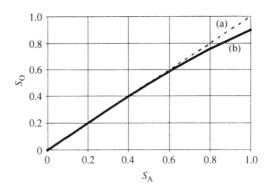

Figure 3.3 Illustration of a typical instrument response.

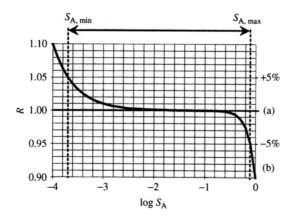

Figure 3.4 Responsivity as a function of input signal.

that has a very small offset (invisible on the scale drawn); this figure also shows a gradual curve that deviates from the straight line at high signal values.

Values of the responsivity, R, for the two lines shown in Figure 3.3 are plotted in Figure 3.4 against the logarithm of the input signal (S_A). The use of the log scale permits us to examine more closely the effect of the offset at very low signal values, without losing sight of the effect of curvature at the high signals. Note that the responsivity for the ideal response has the same value ($= 1.0$) at all signal levels — see line (a) in Figure 3.4.

The responsivity for the 'real' response (line (b) in Figure 3.4) shows two significant deviations from a constant value, as follows:

(i) at *very low* signal levels ($S_A \approx 0$), the *offset* can create significant deviation (equation (3.6));

(ii) at *high* signal levels, the *curvature* of the line also creates a deviation.

The 'linearity' range is a performance characteristic that identifies the range of values of S_A over which the response is 'linear'. This gives an effective operational range for the instrument of ($S_{A,max} - S_{A,min}$).

The problem lies in defining what is 'linear'. Some degree of non-linearity is acceptable, with the range of 'linearity' of a system being normally expressed as the range of signal values (S_A) over which the responsivity (R) is *sufficiently* constant. In practice, this often means the range over which R does not change by more than 5% from the nominal value, as shown in Figure 3.4.

For modern systems, the *offset error* is normally very small and its effect on the linearity only occurs at very low signal levels. Hence, the deciding parameter for the linearity range is normally the point, at *high signal levels*, at which the response changes by 5%. For example, the range of a spectrophotometer is often expressed as the maximum absorbance value up to which the system remains 'linear'.

DQ 3.4

The set of data in the following table gives the input, S_A (gC/s), and output, S_O (A), signals for a flame-ionization detector (FID)-GC system:

Parameter	Value			
	1	2	3	4
S_A	1.0×10^{-12}	1.0×10^{-11}	1.0×10^{-10}	1.0×10^{-9}
S_O	1.50×10^{-14}	1.05×10^{-13}	1.00×10^{-12}	9.99×10^{-12}
R				
Parameter	5	6	7	8
S_A	1.0×10^{-8}	1.0×10^{-7}	1.0×10^{-6}	1.0×10^{-5}
S_O	9.96×10^{-11}	9.87×10^{-10}	9.60×10^{-9}	8.74×10^{-8}
R				

By using these data, (i) calculate the Responsivity, R, at each of the input signal levels, and (ii) hence estimate the Linearity Range, assuming limits of $\pm 5\%$.

Answer

The responsivity is simply the ratio of the signals, S_O/S_A, and gives a maximum value of 0.015 for the smallest signal, 0.00874 for the largest signal, and a value close to 0.01 C/gC† for the mid-range signals. The upper and lower limits for a linearity range within $\pm 5\%$ will be given by 0.0105 and 0.0095 C/gC, respectively. These occur at signal levels of 1.0×10^{-11} and approximately 1.0×10^{-6} gC/s, respectively, giving a linearity range (ratio) of 10^{+5}.

The *linear dynamic range* is the *ratio* which compares the largest signal, $S_{A,max}$, and the smallest signal, $S_{A,min}$, within the linear range that can be measured with acceptable accuracy. The dynamic range of the input signal (DR_A) is given by the following:

$$DR_A = S_{A,max}/S_{A,min} \tag{3.8}$$

The linear dynamic range is an important factor in instruments that are expected to operate at both *trace* and *assay* levels of analysis, i.e. a large ratio of possible

---

† Units are expressed here as coulombs (C) per grams (carbon) (gC).

concentration levels. A large linear dynamic range reduces the need to dilute the more concentrated samples.

DQ 3.5

Reducing the value of minimum detectable signal, $S_{A,min}$, by a **factor of two** will have the following effects:

(a) increase the linear range by a factor of about two (True/False?);

(b) increase the linear dynamic range by a factor of about two (True/False?).

Answer

*The linear range is given by the difference, $S_{A,max} - S_{A,min}$. However, $S_{A,min}$ will be small and its change will have little effect on the value of this range (a, false). However, the dynamic range is given by the **ratio**, $S_{A,max}/S_{A,min}$, where the fractional change in $S_{A,min}$ will have the same fractional change in the **ratio** (b, true).*

3.2.5 Instrument Selectivity (Specificity)

The distinction between *method* selectivity and *instrument* selectivity was preciously developed in Section 1.5.3. An instrument may be *specific* (high selectivity) as a result of design, e.g. by using a detector that has minimal responsivity to the unwanted analyte, as follows:

$$\text{Selectivity} = \frac{\text{Responsivity to Wanted Analyte}}{\text{Responsivity to Unwanted Analyte}} \qquad (3.9)$$

For example, many GC detectors are specific to particular analyte groups; in this case, if it is required to measure a different analyte, then it may be necessary to use a different detector.

Alternatively, *selectivity* can be achieved if different *experimental conditions* in the instrument lead to different analyte sensitivities. Obvious examples of 'internal' selectivity include the following:

• chromatography, which separates analyte components *in time*;

• spectrophotometry, which separates on the basis of response to *different wavelengths*.

DQ 3.6

Which of the following characteristics of a monochromator (see Chapter 16) is a measure of its selectivity:

(a) stray light;
(b) spectral bandwidth;
(c) peak transmittance?

Answer

*The selectivity of a monochromator is based on its ability to differentiate between wavelengths, i.e. to transmit one wavelength while absorbing nearby wavelengths. Spectral bandwidth (b) is the characteristic that describes the **range** of wavelengths that are simultaneously transmitted, and a narrow spectral bandwidth will give high selectivity.*

3.3 Detectability Characteristics

When is detectability a useful performance characteristic?

The instrumental measurement of low signal levels is usually limited by the variation in results that occurs between successive analyses, i.e. method repeatability. The various definitions of detectability are based on the repeatability standard deviation, σ_S, of the output signal, S_O. This value can be calculated [1] from the standard deviation of replicate measurements made by using either blanks or samples with concentrations close to the detection limit.

There are three main detectability levels commonly quoted and these are described in Table 3.1.

DQ 3.7

In a *flame* (F)-atomic absorption spectroscopy (AAS) instrument, I can measure the short-term noise in the output. Can I use the standard deviation of this noise as my estimate for σ_S when calculating the *MDQ* (Yes/No?).

Answer

*No — the measurement of repeatability requires that the measurement **procedure** be repeated in order to identify all sources of variation that may occur between the measurement of the blank and the sample, e.g. drift in the output due to variations in the flame conditions, or changes in the background due to differences in the sample matrix. However, for a system such as a **chromatograph**, the observation of a signal peak relies on its differentiation from the 'blank' baseline signal. In this case, the noise in the baseline can be used.*

The limits are sometimes expressed by using an expression for the minimum detectable *signal*, as follows:

$$S_{O,min} = S_{blank} + (3 \times \sigma_S) \tag{3.12}$$

There can be some confusion between the different expressions, but this can be resolved by reference to the calibration graph represented in Figure 3.5. In equation (3.10), the *MDQ* is expressed in terms of the actual analyte quantity,

Table 3.1 Characteristics of the three main detectability levels

Detectability level	Characteristics
Minimum Detectable Quantity (*MDQ*) or Limit of Detection (*LOD*).	Lowest signal level that can be confidently distinguished from repeatability variations: $$MDQ = (3 \times \sigma_S)/R' \qquad (3.10)$$ where R' is the slope of the calibration curve of the signal, S_O, against the analyte quantity, S_A. Other multiplying constants are sometimes used (e.g. $((3.3 \times \sigma_S)/R')$. In the water industry, the use of a constant of 4.65 is based on the estimation of repeatability obtained from the actual experimental data
Limit of Quantitation (*LOQ*)	Minimum signal level at which the statistical error due to the variations is less than 10%: $$LOQ = (10 \times \sigma_S)/R' \qquad (3.11)$$
Reporting Limit	Used in an impurity audit (e.g. in drinking water) to define the level below which the impurity is not reported, even if it is observed. This will depend on the specifications for the audit as well as the statistics of measurement

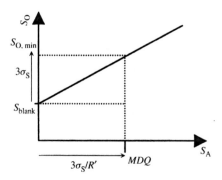

Figure 3.5 Calibration graph for the detectability characteristics; R' is the slope of the line.

S_A, while in equation (3.12) it is expressed in terms of the output signal, S_O, from the instrument.

The value of σ_S includes an estimation of the standard deviation of the 'blank' signal. It is important to realize that there are two factors that contribute to calculating this standard deviation, as follows:

- measurement imprecision due to the *analytical method*;
- imprecision in *producing* a reproducible blank that is truly representative of the sample matrix.

DQ 3.8

The *MDQ* limits are standard specifications for F-AAS, but not for graphite furnace (GF)-AAS. Is this due to problems with:

(a) repeatability of the GF-AAS measurements;
(b) furnace-temperature repeatability;
(c) blank reproducibility;
(d) sample introduction to the furnace?

Answer

The measurement of repeatability should include variations due to all of the listed factors, and these effects should all be quantified in the estimation of σ_S. However, GF-AAS is particularly sensitive to interferences arising from the sample matrix (see Section 6.3). Such interferences are normally counteracted by the use of blanks, but for GF-AAS it is more difficult to produce representative and reproducible blanks than it is for F-AAS. This problem of quantifying the blank response limits the accuracy of some trace measurements.

It must be stressed that detectability figures should be used only to identify the range of signal levels that should be *avoided* if possible, and the analytical process should be developed, as far as possible, to keep the signal levels well clear of these limiting values. The usefulness of such figures is more an indication of the precision of the measurement rather than an invitation to work to these limits.

The term 'sensitivity' is sometimes used as an expression of minimum detectable quantity. However, as 'sensitivity' is sometimes regarded as being equivalent to responsivity, it is clear that its use can lead to possible confusion.

SAQ 3.3

The responsivity of a particular flame-ionization detector is 2.2×10^{-2} C/g (for toluene), with the limit of detection being quoted as 1.5×10^{-12} g/s.

(i) What is the form of the output signal of the detector (e.g. voltage, current, etc.)?

(ii) Estimate the output signal which is equivalent to the Limit of Detection (*LOD*)?

(iii) What is the Limit of Quantitation (*LOQ*)?

3.4 Interaction between Characteristics

This is more complex than it seems at first!

One of the reasons for knowing the performance characteristics of the instrument is to be able to estimate the errors in the *final analytical result*. The uncertainty in the final result caused by instrumental uncertainties will often depend on the *interaction* between different 'types' of characteristics. For this, it is essential to know, not only the Type II and Type II(u) response characteristics, but also the experimental conditions given by the Type I and Type I(u) characteristics.

Figure 3.6 illustrates how a wavelength error (Type I(u)) can generate an error in the observed absorbance (Type II(u)) in a UV–visible spectrum. When making a measurement on the side of the peak, a slight error in the wavelength setting may cause a large error in the observed absorption. The error in the wavelength has a 'first-order' effect on the measured value of absorption. However, near the top of the peak, where the absorption has only a slow variation with wavelength, any error in wavelength will only create a small, 'second-order', effect on the value of absorption.

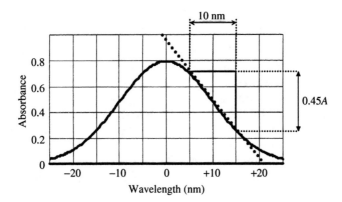

Figure 3.6 Effect of wavelength uncertainty on the observed absorbance in a UV–visible spectrum.

DQ 3.9

A spectrophotometric absorbance line has a peak value of 0.8*A* and a standard deviation in wavelength of 10 nm. Using the data from Figure 3.6, which of the following uncertainties will contribute the greatest uncertainty to the measurement of absorbance at a point 10 nm away from the peak:

(a) a photometric uncertainty of $\pm 0.01A$;
(b) a wavelength uncertainty of ± 0.5 nm?

Answer

From Figure 3.6, the tangent to the spectral line shows that a wavelength 'error' of 10 nm would give an absorbance 'error' of 0.45A. Hence, a wavelength error of 0.5 nm will give an absorbance error of 0.45 × 0.5/10 = 0.0225A. The uncertainty due to the wavelength will, in this case, give an uncertainty in absorbance greater than that due to the direct photometric uncertainty.

It is also possible for the *combined* effect of two uncertainties acting together to be greater than the sum of their individual effects. This can occur if there is a non-linear response in the system. A good example occurs in capillary electrophoresis, where the performance of the system is very sensitive to changes in heat dissipation — see DQ 1.9.

3.5 Memory Effects

How can an instrument have a memory — other than in its computer?

There are a number of examples where material introduced for one measurement 'run' is retained in the instrument system, and appears as 'ghost lines' or 'peaks' in a subsequent measurement.

This is possible in chromatography, where a component with a very long retention time may still be held on the column when a run is terminated, and which will then subsequently appear in a later run as an additional 'ghost peak'.

Automated sample-injection systems (e.g. in chromatography) can also introduce a memory effect through 'carry-over' (see Section 10.2). This occurs when the needle, which draws samples from the vials on a carousel, is not sufficiently well rinsed between operations, and thus allows a fraction of a previous sample to be introduced with the current sample.

There is also a significant problem in atomic absorption spectroscopy where modern instruments perform analyses over a very wide dynamic range of concentration levels, and for many different elements. It is possible for some materials (particularly the refractory elements) to be retained on the burner, furnace or plasma torch. Subsequent measurements are likely to include some of this retained material, which would significantly affect trace measurements.

3.6 Specifications

When buying an item of equipment, why is it more difficult to specify the required performance for an analytical *instrument than it is for a simple* measuring *instrument?*

The experimental measurement conditions of a basic measuring instrument, such as a thermometer, are fixed, and it is usually possible to define the required response unambiguously. For example, the performance of a thermometer can be specified relatively easily by a brief specification, as follows:

Measurement Range $- 20$ to $+ 200°C$;
Accuracy $\pm 0.1°C$.

For the analytical instrument, the experimental conditions of the measurement process can be altered by the operator, and the response of the system will vary depending on these different conditions. Thus, the total specification of performance for the analytical instrument would require the response to be defined under *all possible* experimental conditions. It is this complexity of the possible operating conditions for analytical instruments that makes their performance difficult to define. For example, in Section 5.7 we shall investigate the variations of fractional uncertainty in absorbance, $u(A)/A$, as a function of A for a spectrophotometer.

DQ 3.10

Is the photometric accuracy of a simple spectrophotometer, when it is measuring a sample with an absorbance of $2.0A$ at a wavelength of 360 nm, as good as when it is measuring a sample with an absorbance of $0.8A$ at a wavelength of 560 nm? The answer is no, but why is this so?

Answer

The source used in a simple visible spectrophotometer will normally be a tungsten (or tungsten–halogen) lamp, and we know that the output from this lamp drops very quickly as the wavelength approaches 320 nm (see Section 17.2). With the drop in source intensity, the signal-to-noise ratio in the measurements will become considerably worse. In addition we will see in Section 5.7.5 that the fractional uncertainty in absorbance will be less at 0.8A than at 2.0A.

Instrument specifications, as normally provided by the manufacturer, give information about a selected range of performance characteristics for that particular instrument. The manufacturer will choose to include those characteristics, under specific experimental conditions, that present the instrument in the most favourable light, and may perhaps omit information that compares unfavourably with other instruments in the same price range.

SAQ 3.4

Identify ways in which the analytical **method** can be *designed* to compensate for each of the following listed uncertainties in the performance characteristics (i.e. Types I(u) and II(u) above) of an **instrument**:

  (i) drift in the '0*A*' zero absorbance setting for a single-beam spectrophotometer;
 (ii) imprecision in the injection volumes in a GC autosampler;
(iii) poor output linearity;
(iv) wavelength imprecision in a spectrophotometer.

SAQ 3.5

Explain, with examples, why some instruments quote *detection limits* (in some form) in their specifications, while others do not.

Note. Further questions that relate to the topics covered in this chapter can be found in SAQs 5.2, 6.3, 9.1 and 10.4, and DQ 10.7.

Summary

This *core* chapter has introduced the performance characteristics of the analytical *instrument*, based on the fact that an analytical instrument performs an *experiment*. The performance of the experiment was defined in terms of the conditions of that experiment and the associated instrumental response, together with the uncertainties in both of these.

The *generic characteristics*, common to most analytical and measurement systems, were considered, namely *responsivity, linearity, offset, drift, dynamic range* and *selectivity*. The specific generic concept of *noise*, however, is treated in depth in Chapter 14. The characteristics that are specific to particular types of instrumentation are also discussed elsewhere in chapters dealing with these separate analytical techniques.

Different forms of *detectability* characteristics, which quantify the ability of the instrument to measure trace quantities, were also introduced and compared.

Finally, some *indirect* effects were introduced, including interactions between different characteristics, and the possibility of memory effects between different experimental runs.

Reference

1. 'Guidance for Industry, QB2 Validation of Analytical Procedures: Methodology', *Proceedings of the International Conference on Harmonization of Technical Requirements for Registration of Pharmaceuticals for Human Use*, ICH-Q2B, Section VI, 1996. [Copies available from Drug Information Branch (HFD-210), Center for Drug Evaluation and Research (CDER), 500 Fishers Lane, Rockville, MD 20857, USA, *or* http://www.fda.gov/cder/guidelines/index.htm.]

Chapter 4

Quality Systems in Analytical Measurements

Learning Objectives

- To describe the operational elements of a quality system.
- To discuss the value of a quality system to an analytical laboratory.
- To appreciate the use of 'top-down' and 'bottom-up' approaches in predicting the future behaviour of a system.
- To identify the main approaches being used to develop quality in analytical science.
- To identify the different 'quality' reasons underlying performance-testing protocols.

4.1 Introduction

The assurance of quality in analytical measurements is the driving force behind several current initiatives that are addressing different aspects of the overall problem. This present chapter seeks to outline the framework of current developments in 'quality' that link with the need to understand and quantify the performance of instruments. It aims to identify how these initiatives are clarifying what analysts should know about the instruments that they use.

The role of quality assurance is to ensure and demonstrate that the instrument is, and will continue to be, 'fit for purpose'. In the context of an analytical measurement (see Section 3.1), this means that the instrument should be able to perform *the required analytical 'experiment'* as follows:

- under the required conditions;
- with an appropriate level of response;
- with uncertainties that fall within permitted limits.

4.2 Why is a Quality System Needed?

What is wrong with the way things have been done in the past?

The demands of a modern technological society have highlighted the need for greater assurance in analytical measurements. Obvious examples include the following:

(A) Compliance with specific standards, set by *regulatory authorities*, for measurements made in relation to food, pharmaceuticals, health, etc.

(B) The ability to defend the accuracy of measurements when challenged through *litigation*.

In addition, the new commercial climate is increasingly demanding that service providers (including analytical laboratories) must be able to *demonstrate* that they conduct their business in a manner that *assures* the quality of their service. Large organizations that have themselves adopted a quality system want to ensure that other organizations, which supply them with goods or services, are also operating to similar quality standards.

(C) Analytical laboratories must be able to provide *evidence* of the quality of their service so that they can satisfy the quality demands of their clients.

Interlaboratory comparisons demonstrate that results are not as accurate as many laboratories had previously assumed.

(D) Collaborative trials (Section 4.2.1) have shown that, without joint method development, there are significant differences in the analytical results of different laboratories. The *reproducibility* (Section 1.4.2) between the laboratories can be significantly worse than the *intermediate precision* of the measurement within each laboratory. This evidence shows that significant bias continues to exist between laboratories, and that specific procedures must be developed in order to demonstrate that this bias is within kept defined limits.

DQ 4.1

The need for a quality system is mainly driven by factors from outside the laboratory. Is this statement true or false?

Answer

The immediate response is probably 'yes'. However, the real answer is not so clear cut. The outside factors actually (i) point out that results may not have been as accurate as the analyst thought, and (ii) demand that the analyst is able to demonstrate the claimed accuracy.

Thus, the underlying driving force is that analysts should be confident in the quoted accuracy of their results. This core 'need' should in fact already be an objective of any good laboratory, and, as such, is really a factor that is **internal** to the laboratory.

4.2.1 Collaborative Trials

A typical collaborative trial would involve a number of analytical laboratories analysing separate aliquots of a standard homogenous sample. The results from all of the laboratories are then compared.

Collaborative trials, carried out between reputable and experienced laboratories, have shown considerable variations between the results of the different laboratories, actually exceeding the experimental uncertainties quoted by the laboratories. In particular, the variations are even greater when there is no validated method available for the particular analysis, and where participating laboratories must develop their own procedures.

These trials heighten awareness of the 'hidden' factors that contribute to bias in analysis. In particular, sampling procedures have been identified as giving specific problems, although bias creeps into all aspects of the analytical process.

Once the laboratories are able to compare procedures, it becomes possible to identify the problem areas. Through a process of consultation, based on practical experience, it is possible to develop analytical methods which are 'rugged' (or robust) and translate from one laboratory to another without giving substantially different results. Collaborative trials are now used as a valuable element in the procedure for developing validated analytical methods.

4.3 What is a Quality System?

If I am familiar with a quality system in the production of 'widgets', will I recognize the same elements in a quality system for analytical measurements?

The key elements of *any* quality system are as follows:

> *Objective Experimental Data*, which provides
> *Documentary Evidence*, that enables a
> *Decision* to be made, concerning the
> *Fitness for Purpose* of the item or service, in respect of its
> *Future Intended Use*.

Of these five elements, the one that needs to be stressed is that a *decision* must be made. The decision, although somewhat less dramatic, is similar to that made at the last minute as to whether to launch a space rocket — it is 'Go' or 'No Go'. Either the item or service is performing within the pre-defined levels, or it is not.

The analyst must be able to decide if the system, the equipment, and the method, are 'fit' for the purpose for which it is *going to be used* (note the use of the future tense!). It is not sufficient to know that all went well in the past, and therefore hope that it will be OK in the future. You might agree with this obsevation if you were the actual astronaut on top of the rocket!

A decision must be made, and it is important that this decision should be based on factual, objective, data. Hence, other key elements of a quality system are concerned with the acquisition of these factual data and their proper documentation.

The concept of 'fitness for purpose' is also central to any quality assurance system. The *purpose* must be clearly defined — whether it is the provision of consultancy advice or the analysis of an unknown sample. Criteria must be set at the local level in order to define the laboratory performance that would be sufficient to be *fit* for that purpose.

A quality system takes structured data from past performance in order to enable a decision to be made that the performance has been, and will continue to be, within acceptable limits. The data that are recorded must be carefully documented so that clear decisions can be made. The ultimate purpose of a quality system for an analytical laboratory is therefore to ensure and to demonstrate that the laboratory will continue to meet predefined standards.

The operational aspects that are used to deliver a quality system will include the following:

• Defined standards (external/internal);
• Traceability;
• Quality control;
• Quality assurance;
• Quality management;
• a Quality manual.

These are the same broad elements for *any* quality system, and the objectives of each element will be similar for any system. However, the detail within each element will depend on the purpose of the particular system.

Integral to the idea of 'quality' is the concept of a 'standard' against which outcomes or activities can be judged. The multi-purpose use of the word 'standard' may lead to confusion in relation to quality systems, and in analytical chemistry the term 'standard' appears in respect of at least three different types of situations, as follows:

• Chemical standard, e.g. a sample of known concentration;

• Performance standard, e.g. being able to measure concentration to $\pm 1\%$;

• Quality standard, i.e. adopting recognized methods of organization and working.

It is important to understand that external standards do not necessarily define every detail of local work. Under the relevant International Organization for Standardization (ISO) standard, ISO 9000, for example, it is the responsibility of the particular organization to define the standard of service that it aims to provide. The role of the ISO standard establishes the minimum standards in *quality assurance* by which the organization must be able to demonstrate that it meets its *own* standards of service. The choice of standards depends on the particular circumstances (Section 4.6).

Traceability to standards is an essential concept within any quality system (again, we can see here the use of two meanings for 'standards'). The following are of particular significance:

(A) It is important to be able to trace the lines of responsibility in management and how information flows through the system (*standards* in quality management).

(B) In an analysis, it is also important to trace how the final analytical results can be related to chemical (or physical) certified reference materials (CRMs).

In addition, it is important to establish continuity and traceability through *time*. It must be possible to trace back the steps that led up to a final result, thus establishing accountability for past analytical results. Traceability also links forward to give an assurance about the future operation of the system.

Quality control encompasses those activities, carried out within the system, that are designed to keep the system operating within the accepted standards of service or work set by the organization. Quality control of the analytical process requires the systematic monitoring of method and instrument performance (Section 4.13). Thompson and Wood [1] have identified guidelines for the implementation of an internal quality control (IQC) system in analytical chemistry laboratories.

Quality assurance is that set of activities which confirm that the quality control process is performing adequately and that the system is operating within the accepted quality standard. Quality assurance provides *evidence* of satisfactory operation through careful documentation. For example, Huber [2] has identified the quality assurance issues that arise during the lifetime of an HPLC system.

A quality system will provide the necessary documentary evidence that will demonstrate (to outsiders, if necessary) the quality of the work. However, it will also include internal processes that will help the laboratory to improve performance and learn from previous experience.

The management of everything within the system (staffing, operational procedures, interaction with clients, etc.) must also reflect the needs of the quality system. Without an adequate standard of *quality management*, the other elements of the system can fail to link into a satisfactory whole.

The details of all essential procedures within the quality system must be documented in a *quality manual*. This reference document defines the quality system at all levels, and sets out, in detail, all of the essential procedures. An important benefit of the quality manual is that the ways in which specific tasks are carried out are written down, and are therefore less dependent on the specific knowledge of particular members of staff. This reduces inconsistencies that could arise due to staff changes or absences.

SAQ 4.1

As a means of ensuring traceability of performance, should I test the performance of my spectrophotometer before, or after, a routine maintenance procedure, or both?

4.4 Benefits of a Quality System

Who gets the benefit if my laboratory introduces a quality system?

All quality systems involve significant amounts of documentation, record-keeping, and operational discipline. Where are the advantages of such systems?

A quality system for analytical measurements must satisfy one of the key objectives of the *professional* approach to analytical measurements — *it must provide the analyst with the ability to confirm the validity of the analytical results that are produced.* Flowing from that central objective, a number of secondary advantages can then be identified.

The *external* advantages are as follows:

- the *evidence* of a quality system gives the client of the analytical service confidence in the reliability of the results;
- the client benefits from this improved level of assurance;
- the laboratory will benefit by maintaining and developing its existing client base.

There are significant *internal* advantages as well. These are as follows:

- the analyst benefits through a greater *clarification and quantification* of all of the steps that contribute to the final result;
- the analytical process, from the initial discussions with the client, through sampling and analysis to presentation of the final results, becomes more visible to all those involved and does not depend on the knowledge of specific staff;
- the actual process of *developing* the quality system can also pick up deficiencies in earlier procedures that may have led to errors in the results;
- increased reliability of results reduces the need to rerun tests for confirmation when faced with unexpected results.

DQ 4.2

List the following potential benefits of a quality system in order of priority with respect to *your own* analytical measurements, and explain your reasoning:

(a) meet the requirements of regulatory authorities;
(b) improve the efficiency of your work;
(c) confirm your own professional approach;
(d) prepare in case of litigation.

Answer

Clearly, your detailed answer will depend on your own particular situation with respect to analytical measurements. Nevertheless, whatever your situation, your first priority should be (c). Every responsible person has a quality system of sorts. However, your own ad hoc *quality system may well be missing important links in the quality chain.*

4.5 Top-Down and Bottom-Up

Will it still be 'within calibration' tomorrow?

The previous section stressed the role of a quality system in confirming that an item or service will be fit for its intended purpose. In this section, we will introduce the idea that confirming performance quality requires the mutual support of *two* different confirmation mechanisms, as follows:

- Confirmation by Outcomes (Top-Down);
- Prediction by Structure (Bottom-Up).

As a very simple example, take the case of the production of a wooden school ruler. If this item is to be fit for purpose, it must continue to be capable of measuring distances (up to about 30 cm) with the required degree of accuracy under a variety of environmental conditions.

Before the item (ruler) leaves the factory, there must be a mechanism to ensure that it is calibrated to a sufficient accuracy, i.e. the *outcome* of its use (as a ruler) must be confirmed to an external standard of length. This is a *top-down* confirmation of quality.

Calibration, however, is only truly valid *at the time* of calibration, and does not, of itself, give confirmation of *future* accuracy. If we are to be sure that the ruler will continue to be accurate, we need to be confident that it will not stretch or shrink, either as a result of different environmental conditions (temperature, humidity, etc.), or due to the passage of time. For this, we need to know that the selected wood and the method of manufacture (e.g. grain orientation) will make the calibration of the ruler resistant to such changes. The required *structural* evidence comes from using knowledge of the detailed design, development and manufacture of the item. This structural evidence is *bottom-up* data.

With evidence of the *structure* of the item, *together* with the calibration, we can make reasonable predictions about *future* performance.

The same intermixture of a top-down check on outcomes, and a bottom-up confirmation of structure, occurs in all quality systems. As a very different example, we can look at a quality assurance system for the management of an instrument maintenance department. Its past performance can be checked against a number of specified outcomes (e.g. the down-time of clients' machines). Expectations of future performance can only be made by *also* taking into account such factors as the quality of the organizational structure of the service and its staff training programme.

Figure 4.1 represents the *outcomes* of a complex system 'now' (N) and at some time in the 'future' (F). Some outcomes (S) are tested 'now' against external standards, and the 'future' outcomes are confirmed through knowledge of the structure of the system. Confirming quality requires that it is possible to demonstrate that *every* outcome has met, and will continue to meet, minimum performance standards. However, Figure 4.1 shows that there are also some 'now' outcomes that are not tested against external standards. In fact, it is not necessary, or possible, to test *every* possible outcome. We can use our knowledge of structure (bottom-up) to confirm that all outcomes, that have not been specifically checked, will still be met.

As an example, it is not possible to check the wavelength of an instrument at 'all' wavelengths (e.g. every 1 nm). Instead, we use traceable standards (calibration filters) to confirm the quality of *some* performance outcomes, and use our knowledge of the structure of the item to predict that the performance of all other outcomes will be within their specified limits.

Assuming that we know that (i) the monochromator in a spectrophotometer has been designed and manufactured to acceptable standards (bottom-up), and

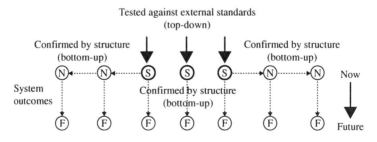

Figure 4.1 Application of the 'top-down' and 'bottom-up' confirmation mechanisms in testing the outcomes of a complex system.

(ii) that it has been confirmed (top-down) to be within tolerances at 300 and 400 nm, then we can be assured that it will not show an unacceptable error at 350 nm.

The time between (top-down) calibration checks depends on what is being checked, and the time it is expected to take to drift out of calibration (Section 4.12).

DQ 4.3

In an experiment to measure the concentration, c_A, of an unknown sample, its absorbance, A_A, is compared to the absorbance, A_S, of a standard sample of known concentration, c_S. The unknown concentration, c_A, is calculated by using the following formula:

$$c_A = c_S \times (A_A/A_S)$$

The absorbance of the standard sample gives 'top-down' calibration data. What 'bottom-up' information must be assumed about the performance of the instrument?

Answer

The use of the single calibration point assumes that the instrument gives a linear (straight-line) response of apparent absorbance versus concentration, and that this line passes through the origin. The expectation of this linear response can be confirmed by a knowledge (bottom-up) of the instrument design.

4.6 Approaches to Quality

If there is no 'single' answer to the question 'what to do about quality?', which 'answer' is the one that is most appropriate to my needs?

It is not possible to define a 'typical' analysis. Analytical chemistry contributes to almost every aspect of our modern society, i.e. food, health, the environment, trading standards, agriculture, etc. Consequently, analyses are performed over wide ranges of many different 'areas', including the following:

• types of analytes;
• concentration levels (from trace impurities to assay);
• sample states;
• interfering species.

In addition, varying requirements mean that analyses are performed (i) to different levels of accuracy, and (ii) to satisfy different forms of regulatory compliance.

The variety of possible analyses makes the establishment of an all-embracing quality system difficult. Nevertheless, much work is being done to develop quality through several different *approaches* to the problem. Each approach has been developed in order to satisfy particular needs. The main approaches that are relevant to analytical chemists are as follows:

- Quality Standards and Accreditation (Section 4.7);
- Valid Analytical Measurement (VAM) (Section 4.8);
- Proficiency Testing (Section 4.9);
- Certified Reference Materials (CRMs) (Section 4.9);
- Validated Methods (Section 4.10);
- Collaborative Trials (Section 4.2.1);
- System Suitability Testing (SST) (Section 4.11);
- Quality Control of Instrument Performance (Section 4.13);
- Equipment Qualification (EQ) (Section 4.12).

We saw above in Section 4.5 the idea that confirming 'fitness for purpose' requires a mixture of top-down and bottom-up processes. The above list reflects (very approximately) where different initiatives stand in a 'top-down' hierarchy. These are discussed separately in the following sections.

The top-down process is related to analytical objectives, and can be considered [3] to be *validation*. The bottom-up process confirms the 'credentials' of the analytical instrument to carry out the analysis, and can therefore be called *qualification*.

4.7 Quality Standards and Accreditation

Can there be a standard without accreditation, or vice versa?

Quality standards must, themselves, be 'fit for purpose', and a variety of published standards exist which cover many purposes. The standards and guidelines that are most applicable to the context of analytical instrumentation are as follows:

- The ISO 9000 Series — general quality standard for industry.
- The Good Laboratory Practice (GLP) scheme — developed for use by regulatory bodies.
- ISO Guide 25 — general requirements for the competence of calibration and testing laboratories.

With increasing international consensus on the need for quality standards, there are national and international equivalents to these standards, e.g. the European Committee for Standardization (CEN) and the Organization for Economic Cooperation and Development (OECD) standards (Europe), and British Standards (BSs) (UK).

Accreditation is a process whereby a recognized independent body will confirm, and provide certification, that the laboratory carries out defined work according to the rules of a given quality standard. This may occur at different levels — either in relation to a general field of activity or in relation to specific tasks.

Accreditation to ISO 9000 standards confirms that a company has established acceptable operating procedures, *throughout its organization*, to meet its *own defined standards*, e.g. in handling customer complaints, managing staff training, and calibrating instruments. Accreditation from the UK National Measurement Accreditation Service (NAMAS) would, however, confirm that a laboratory has demonstrated a capability to perform *specific* tests to *externally defined* levels of accuracy. The NAMAS accreditation would only apply to the *defined range* of tests, and not to other types of tests that may be carried out by that laboratory.

DQ 4.4

'Measurements carried out within *accredited laboratories* must be made to an accuracy of at least 1.0%'. Is this statement true or false?

Answer

This statement is too ambiguous to be either true or false — 'accreditation' can be given for a wide range of different standards. For example, accreditation to ISO 9000 establishes an overall quality management standard, with the accuracy targets being set locally (not necessarily to 1%). Accreditation to the United Kingdom Accreditation Service (UKAS) (i.e. NAMAS) standards might only apply to a specific subset of measurements within the laboratory, and the level accuracy will be defined by the specific test being used.

An *international* or *national* quality standard must display the following features:

- a wide applicability for laboratories carrying out different types of analyses, for different clients, and with different types of instruments;
- the ability to permit local requirements to dictate the details of local standards.

The quality standard will define those elements of a system that are indeed common across all relevant sectors, but must also allow for detailed performance standards to be developed locally for the specific purposes. It is therefore necessary to *interpret* the broad objectives of international standards when applied to particular situations.

The interpretation of a quality standard is very much a 'top-down' process. The top 'layer' sets out the general principles about *managing* quality, with the succeeding layers becoming increasingly more specific about particular applications.

The ISO 9000 Series is a set of flexible quality standards that can be applied to companies, both large and small, engaged in a variety of different fields.

These standards establish the framework within which the *quality system* should operate, but do not tell the company how they should conduct their core business. It is for the company itself to define their own local objectives and performance standards within the ISO 9000 quality framework.

Many companies who have achieved the ISO 9000 standards now demand that their suppliers of goods and services should also be able to demonstrate quality to these same standards.

The GLP scheme was developed in 1976 by the Food and Drug Administration (FDA) as a set of principles for work carried out for regulatory authorities in the USA. In 1982, the OECD published an international standard for GLP so that work could be recognized between the countries using these principles.

The ISO *Guides* seek to develop appropriate interpretations in selected areas, and are a useful mechanism by which industry can develop sets of local standards that still have international recognition. These guides still require interpretation in terms of detailed application, and the process of Equipment Qualification has been developed as an implementation of part of the ISO Guide 25.

4.8 Valid Analytical Measurement (VAM) Programme

The Valid Analytical Measurement (VAM) Programme [4], funded by the UK Department of Trade and Industry (DTI) and administered under contract by the LGC (formerly The Laboratory of the Government Chemist), provides support for analytical laboratories that wish to develop a quality approach to analytical measurements.

The programme has identified the following six principles that need to be addressed when providing an analytical service:

(1) Analytical measurements should be made to satisfy an agreed requirement.
(2) Analytical measurements should be made by using methods and equipment which have been tested to ensure that they are fit for their purpose.
(3) Staff making analytical measurements should be both qualified and competent to undertake the task.
(4) There should be a regular independent assessment of the technical performance of a laboratory.
(5) Analytical measurements made in one location should be consistent with those made elsewhere.
(6) Organizations making analytical measurements should have well-defined quality control and quality assurance procedures.

These principles are designed to ensure that the analytical service is 'fit for purpose' by ensuring that the purpose is well defined and agreed, and that the analysis is of sufficient accuracy and does indeed match the agreed purpose.

SAQ 4.2

Why is the first VAM principle, i.e. 'Analytical measurements should be made to satisfy an agreed requirement', an important part of a quality system for analytical measurements?

4.9 Proficiency Testing and Certified Reference Materials

How are proficiency testing schemes and certified reference materials related?

Proficiency testing (PT) schemes and certified reference materials (CRMs) are both used to establish the accuracy of a laboratory's work by comparison with an external reference.

In a PT scheme, samples of known characteristics are distributed on a regular basis to the participating laboratories, who then analyse the samples by using whatever method they choose as being most appropriate. Comparison between the performance of different laboratories then provides feedback assurance about the continuing performance of *individual* laboratories.

An alternative (or additional) method of external reference is to use CRMs. The establishment of reference standards for physical quantities is relatively straightforward, e.g. for mass and optical transmittance. However, standards for analytical chemistry present a much greater problem due to the great variety (Section 4.6) of the different types of analyses being undertaken.

A CRM must be a pure, stable, standard sample of the required analyte within a matrix that is closely representative of that occurring in the test sample. Currently, a large number of new CRMs are being developed and evaluated, thus increasing the range now available to the analyst.

4.10 Validated Methods

Are standard operating procedures the same as validated methods?

It is essential, for a quality analytical service, that the analytical method is itself 'fit for purpose' — it must give a satisfactory performance when used with the available equipment. Many standard validated methods are documented in various publications, such as the British Pharmacopoeia (BP), the United States Pharmacopoeia (USP), specialist journals, etc. However, new validated methods are continually required [5] in order to take advantage of the recent developments in analytical instrumentation.

Method validation is the *process* [6] that gives the assurance, with documented experimental evidence, that the method being used will indeed carry out the analysis to the expected level of performance. The first step is to define the expected 'performance' (Section 1.5) of the analytical method. The chosen method must fit the requirements, and the state and concentration of analyte to be measured must fall within the accepted range of the method.

The choice of the most appropriate method will depend on a balance between the various characteristics that provide the best 'fit' for the intended purpose. For example, if the minimum accuracy required for the purpose is $\pm 10\%$, there is no point in choosing a lengthy and expensive method to achieve an accuracy of $\pm 1\%$ when a quicker, simple, method gives the result to $\pm 5\%$.

The processes of method *development* and method *validation* are often intertwined. Method development involves the evolution of an analytical procedure that will satisfy the initial objectives. Some of the objective tests carried out in the development stage can be used to provide the necessary factual evidence for method validation at a later stage.

Methods, which are to be accepted as standard methods, will have been subject to collaborative studies involving several laboratories where the 'ruggedness' of the method will be fully tested. Collaborative trials (Section 4.2.1) between laboratories are used to identify and eliminate factors in the method that might be subject to laboratory bias. These tests are essential to improve the ruggedness and reproducibility of the final method. However, a specific laboratory may develop validated methods for its own use. In this case, the ability of the method to be transferred to other laboratories (ruggedness or robustness) will not have been tested.

The EURACHEM guide, 'The Fitness for Purpose of Analytical Methods' [7], reviews the practical steps appropriate to method validation, particularly for those laboratories which are unable to join collaborative trials.

Standard Operating Procedures (SOPs) are part of a quality system, and define the way in which specific procedures must be carried out. Indeed, the standard procedure used for a particular analysis may also be a standard validated analytical method. However, SOPs may also be defined for other routine operational (non-analytical) aspects of the system, e.g. cleaning or calibrating the instrument.

4.11 System Suitability Testing

Suitable for what?

All quality systems which are appropriate to analytical chemistry require that the equipment used is 'fit for purpose'. The System Suitability and Equipment Qualification (see Section 4.12) concepts both seek to address this problem, but in different ways and with different terms of reference.

System Suitability Testing (SST) is essentially a component part of a validated method. It requires that the performance characteristics of the measurement system meet the requirements of the analytical method, and is a 'top-down' check on those performance characteristics that are relevant to the *particular* analytical method under consideration. The instrument system must have an appropriate performance for the given task.

In this context, the 'system' includes *all equipment* used in the process, not just the core analytical instrument, i.e. it includes the glassware, analytical column, balance, etc.

SST methods should check all of those performance characteristics of the instrument system that materially affect the characteristics of the method. For example, *precision* may be the most significant instrument characteristic in determining the *accuracy* of a comparative analytical method. This relationship between method characteristics and instrument characteristics are explored within different contexts elsewhere (see Sections 5.5.1 and 8.7), with the specific tests appropriate to different instrument systems appearing in the relevant chapters describing the various instruments.

4.12 Equipment Qualification

How can I demonstrate that my instrument will continue to be 'fit for purpose' if I don't have the details about how it is designed, how it was constructed, how the software works, etc.?

All quality systems require that the user of the equipment should be able to demonstrate that the equipment is 'fit for the purpose intended', and it is the *responsibility* of the *user* to provide this evidence. Equipment Qualification (EQ) is a standard approach that enables the manufacturer to support the user in confirming that the instrument is indeed fit for purpose.

We have seen elsewhere (Section 4.5) that the demonstration of quality requires both a 'bottom-up' confirmation of *structure* and a 'top-down' confirmation of *performance*. The user of an instrument can indeed confirm *performance* through appropriate calibration and performance checks, but is unable to confirm the *structure* of the instrument without information from the manufacturer.

This is an important point that is worth stressing. Confirmation of the *structure* of an analytical instrument can only be achieved through 'bottom-up' information concerning the design concepts embodied in the instrument (including software), the quality of the components used, and the quality of the manufacturing process. Unless the necessary information can be supplied directly by the manufacturer, the user will have to carry out alternative extensive tests to confirm equipment quality and continuing performance.

The EQ process has been developed to integrate the 'bottom-up' information with the 'top-down' data, and hence confirm that the instrument is indeed 'fit for the purpose intended'. The overall process consists of the following four stages:

- Design Qualification (DQ) (including a User Requirement Specification (URS));
- Installation Qualification (IQ);
- Operational Qualification (OQ);
- Performance Qualification (PQ).

Various publications [3, 5, 8–10] identify the requirements for EQ at the particular stages in the life of the instrument.

The user starts the process, before actually buying the instrument, by defining the minimum levels of performance that the instrument must *continue to meet* in order to carry out its intended tasks. These criteria for the minimum, *ongoing*, 'operational' performance characteristics form the *User Requirement Specifications* (URSs) [3] for the instrument.

The process of equipment qualification then seeks to confirm that the instrument will continue to meet the performance criteria established in these user requirement specifications.

Instrument performance will deteriorate over time due to wear and ageing of the components, and although routine maintenance will counteract this drop in the short term, there will be an inevitable long-term decline. Nevertheless, the performance of the instrument must continue to meet the *minimum* criteria established in the URS. Therefore, in order to allow for this subsequent decline, the *purchase specification* of any instrument purchased should exceed the requirements for the URS by a suitable margin. For example, a new spectrophotometer may have a *specified* photometric accuracy that is three times the minimum required by the URS.

If the operational performance characteristics established in the URS are unnecessarily high, then it will be difficult to maintain the instrument 'within specification'. Huber and Welebob [11] argue that the performance criteria should be no higher than is *appropriate* for the actual needs of the laboratory, and give practical examples of selecting suitable operational characteristics. If, for example, a laboratory carrying out routine assays has more that one instrument of the same type, but which are of different ages and from different manufacturers, the performance testing processes could be simplified by choosing less stringent common operational criteria that all of the instruments can meet. It may be necessary, however, to identify separate instruments for more accurate work, e.g. impurity testing.

The first Equipment Qualification stage, namely Design Qualification (DQ), provides the evidence that the design and construction of the instrument will fulfil the requirements specified by the user in the URS. It will include details of the following:

- functional and operational specifications;
- design development processes;
- quality control used in the manufacture of the instrument.

The DQ information therefore confirms the *design*, *performance* and *structure* of the instrument.

When the instrument is installed, the Installation Qualification (IQ) confirms that the environment (temperature, available services, location, etc.), in which the instrument is to be used, meets the specified requirements. The IQ then confirms that the instrument is complete as ordered, has been correctly installed in the appropriate environment, and is seen to be operational when switched on.

The next stage, namely Operational Qualification (OQ), is an *ongoing* process of 'pass/fail' performance testing that confirms that the instrument continues to work to the operational criteria set out in the URS. An initial OQ test will normally be carried out as soon as the instrument has been correctly installed, i.e. following the IQ. Subsequent OQ testing will then be carried out as follows:

- on a routine basis to confirm ongoing performance;
- after events which may affect performance, e.g. component replacement.

The exact sequencing of OQ testing and maintenance actions will depend on whether the testing is intended to confirm performance up to the time of the maintenance, and/or confirm performance after the maintenance and into the future.

Performance Qualification (PQ) is also an *ongoing* process of 'pass/fail' performance testing, but is used in order to confirm that the instrument is working to those aspects of its specification that are *appropriate* for the analytical methods that are *routinely used* in the particular laboratory. PQ involves testing performance under conditions equivalent to those used in actual analytical methods.

Performance Qualification (PQ) is the culmination of a series of equipment qualification stages that trace the performance of an instrument from its original design through a *bottom-up* approach, and checks that the *specific analytical* instrument is within the performance requirements for its *general laboratory use*. This can be compared to System Suitability Testing (SST), which checks *all equipment* used in the process but only for the performance characteristics required for the *particular analytical method*. The two tests have common elements, but each may also include some additional elements that are not included in the other.

The manufacturer will normally deal with the following:

- supply DQ information on the instrument performance, design and construction;
- install the instrument, and carry out an IQ test;

- perform an initial OQ test to confirm that the instrument is performing to specification;
- carry out an initial PQ test by using a standard set of operating conditions and samples (supplied by the manufacturer).

Many instruments have the option of automatically performing, under software control, many of the recognized tests for OQ, PQ and SST. Some of these tests might require the use of traceable standards (e.g. calibrated filters) which may be held within the instrument itself.

Both OQ and PQ require ongoing performance testing of the instrument and might appear to overlap. Indeed, some of the tests may be identical, but it is useful to stress their different objectives, as follows:

- OQ tests establish the traceability of instrument performance back to its intended specifications;
- PQ tests confirm the performance as part of an analytical requirement, and, through SST, may be part of the method validation process.

The link through OQ and PQ to SST acts as the link between Equipment Qualification (bottom-up qualification) and Method Validation (top-down process) in respect of equipment performance.

The practical differentiation between OQ and PQ can be seen with the example of modular HPLC systems (see Section 9.4). However, this difference is not so clear when the instrument is a single integrated system, e.g. a UV–visible spectrophotometer (see Section 5.5).

DQ 4.5

Identify which of the following might be true.

The period between testing for Operational Qualification (OQ) is:

(a) defined for each type of test (e.g. every month for wavelength accuracy);
(b) determined by cost efficiency;
(c) defined by the type of instrument;
(d) planned within the particular EQ process;
(e) constrained by the User Requirement Specification (URS);
(f) defined by the application (e.g. daily for a drug analysis).

Answer

The period between tests should be such that the instrument being tested will not normally drift into a 'fail' condition. If the period is too great, and an instrument fails a test, then the traceability of performance backwards is lost, and it is not possible to be sure for what previous period the results were, or were not, valid. If the period is too short, then time and money is wasted on excessive repeat testing. The time taken to drift 'out of spec' will depend on the tightness of the URS in relation to the instrument design specifications.

The true responses are therefore (b), (d) and (e).

External constraints (e.g. regulatory compliance) will often define testing for applications (e.g. SST). However, this is not part of the OQ process.

It is only through experience that it is possible to set the most appropriate levels of *acceptance criteria* (for OQ and PQ) that will provide the following:

- confirm satisfactory analytical performance;
- allow sufficient time between tests so that the process is not unduly onerous.

In summary, Equipment Qualification is the responsibility of the *user* of the instrument. It is therefore very important that the initial URS and the subsequent EQ process match the needs of that particular user. Some manufacturers are now willing to modify their standard OQ and PQ procedures, so that they can set up a testing routine which does match the specific needs of the customer.

SAQ 4.3

Differentiate between System Suitability Testing and Equipment Qualification on the basis of 'top-down' and 'bottom-up' approaches to quality assurance.

4.13 Quality Control of Instrument Performance

I seem to be carrying out the same performance test for different reasons.

Quality control (QC) is the generic name that includes all those actions taken to assure general quality. The internal quality control of analytical data (IQCAD) [12] covers those actions (other than sampling) which are appropriate to the instrumental measurement process.

An important aspect of the quality control of an instrument is the continuous monitoring and recording of performance data. This enables trends in performance to be picked up early, and provides a quantifiable process that warns of potential 'out-of-control' situations. Shewhart and CUSUM charts [13] provide warning and action limits based on the historical sequence of performance characteristics. This historical information is particularly helpful in establishing maintenance and repair schedules for the instrument.

For any instrument, therefore, there is a range of different performance tests that might be required for different purposes, as follows:

- OQ — to confirm the performance of the instrument as specified;
- PQ — to confirm the performance for its routine use in the laboratory;
- SST — to confirm the performance for specific analytical methods;
- IQCAD — to monitor performance in order to facilitate effective maintenance.

It makes sense, with a number of similar instruments in the laboratory, to review what tests are required, and how often. By careful planning, it is possible to avoid duplication of effort, and to allow each test to provide data for more than one objective.

SAQ 4.4

You check the performance of an analytical instrument after some time in use, and find that it falls outside the manufacturer's specification. What do you do?

SAQ 4.5

A performance check may be carried out for a variety of different 'quality' reasons (e.g. OQ, PQ, SST and IQCAD). Identify the possible reasons for which a check on each of the following might be carried out:

(i) The noise level in an FTIR spectrophotometer;
(ii) The total energy at the detector in an FTIR spectrophotometer.

Note. Further questions that relate to the topics covered in this chapter can be found in DQs 1.6 and 11.1.

Summary

This chapter has set the 'quality' context in this present text. It has stressed the fact that *quality* should not be seen as an external imposition on the work of the analyst, but should be welcomed as part of the approach of a professional analyst. The concept of a quality *system* was introduced and the benefits and demands of such a system were discussed.

A central theme of the book was developed here through an appreciation of the *mutual support* of both 'top-down' and 'bottom-up' approaches to establishing the functional quality of any item or operation. This theme was used to introduce the variety of different approaches which are currently being used to address quality issues in analytical science.

The roles of Equipment Qualification and System Suitability Testing were examined in relation to the 'bottom-up' *qualification* and 'top-down' *validation* of performance. The various stages of the Equipment Qualification process were discussed in some detail, highlighting the roles of the user and the manufacturer.

References

1. Thompson, M. and Wood, R., *Pure Appl. Chem.*, **67**, 649–666 (1995).
2. Huber, L., *Accredit. Qual. Assur.*, **1**, 24–34 (1996).
3. Bedson, D. and Sargent, M., *Accredit. Qual. Assur.*, **1**, 265–274 (1996).
4. Sargent, M. and Hammond, J., *Spectroscopy*, **12**(7), 46–47 (1997).
5. Hammond, J., *Spectroscopy*, **12**(9), 46–50 (1997).
6. Jenke, D. R., *Instrum. Sci. Technol.*, **25**, 345–359 (1997).
7. EURACHEM, *The Fitness for Purpose of Analytical Methods*, ISBN 0-948926-12-0, 1998. [Copies available from VAM Helpdesk, LGC (Teddington) Ltd, Teddington, UK.]
8. Burgess, C., Jones, D. G. and McDowell, R. D., *Analyst*, **123**, 1879–1886 (1998).
9. Freeman, M., Leng, M., Morrison, D. and Munden, R. P., *Pharm. Technol. Eur.*, 40–46 (November 1995).
10. Huber, L., *Accredit. Qual. Assur.*, **4**, 87–89 (1999).
11. Huber, L. and Welebob, L., *Accredit. Qual. Assur.*, **2**, 316–322 (1997).
12. Analytical Methods Committee, *Analyst*, **120**, 29–34 (1995).
13. Miller, J. C. and Miller, J. N., *Statistics for Analytical Chemistry*, 3rd Edn, Ellis Horwood, Chichester, UK, 1993, pp. 92–98.

Chapter 5

UV–Visible Spectrophotometer Systems

Learning Objectives

- To interpret spectrophotometer specifications in terms of their impact on analytical method characteristics.
- To identify the advantages and disadvantages of single-beam and double-beam systems.
- To demonstrate quantitatively how analytical uncertainty can be a function of the experimental conditions.
- To identify which performance characteristics should be checked to confirm performance quality.

5.1 Basic (Single-Beam) System

The function of a UV–visible spectrophotometric system is to record how much light is absorbed by an analytical sample. The measurement is made at a specific wavelength, λ, selected within the ultraviolet and visible range — approximately 190 to 1000 nm. Such systems are commonly used in stand-alone spectrophotometers, and also as detectors in HPLC systems.

The UV range extends from about 190 to 350 nm. The visible range extends approximately from 350 to 650 nm, although some 'visible' instruments often extend up to about 1000 nm, i.e. into the near-infrared region.

The basic layout of a single-beam UV–visible system is given in Figure 5.1. The light, containing a continuous spread of wavelengths, is provided by an appropriate source, which is usually either a deuterium lamp (for UV light) or

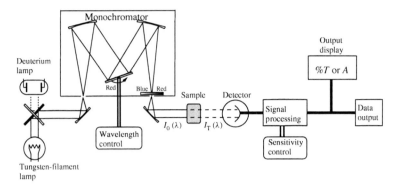

Figure 5.1 Schematic representation of a single-beam UV–visible spectrophotometric system.

a tungsten-filament lamp (for visible light). The source can be selected by a switchable mirror as required.

The next stage, the monochromator, selects the wavelength, λ, at which the measurement takes place. However, it is not possible to select only *one* wavelength, and a narrow band of wavelengths on either side of λ will also be transmitted. The width of this band is called the Spectral Bandwidth, $\Delta\lambda$ (see Section 16.2.2). The (nearly) monochromatic radiation then passes through the sample where some absorption may take place. The intensity of the transmitted radiation, $I_T(\lambda)$, is finally recorded at the detector.

The detector produces an electronic signal (normally proportional to the radiation intensity) which is then amplified and made available for direct display and/or transfer to a microprocessor. A *sensitivity control* allows the amplification of the signal to be controlled, either by an internal microprocessor or by the operator.

DQ 5.1

Could the monochromator be placed between the sample and the detector instead of between the sample and the source?

Answer

For the basic operation of the system, it does not matter where the wavelength selection occurs — either before or after the sample. However, for UV systems it is preferable to place the monochromator before the sample in order to reduce the total amount of UV energy which falls on to the sample. This reduces the stray light that could be re-emitted (possibly at the measurement wavelength) due to fluorescence of the sample.

A typical spectrophotometric measurement measures the concentration of the specific analyte dissolved in a solvent — the *test sample*. A sample of solvent without the analyte is used as the *reference sample*.

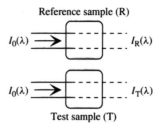

Figure 5.2 Comparison of the light transmitted by the test and reference samples.

The *same* intensity of incident light, $I_0(\lambda)$, falls on both the test sample and reference sample. It is the presence of the analyte that reduces the light transmitted from $I_R(\lambda)$ to $I_T(\lambda)$ (see Figure 5.2). The spectrophotometer compares the light, $I_T(\lambda)$, transmitted by the *test* sample, T, with the light, $I_R(\lambda)$, transmitted by the *reference* sample, R.

The *transmittance*, T_λ, of the *analyte* (for the wavelength λ) is defined by the ratio of the two intensities of the light, i.e. $I_T(\lambda)$ and $I_R(\lambda)$. In most cases, the solvent will have little absorbance, and $I_R(\lambda) \approx I_0(\lambda)$:

Transmittance, T_λ
$$T_\lambda = \frac{I_T(\lambda)}{I_R(\lambda)} \approx \frac{I_T(\lambda)}{I_0(\lambda)} \tag{5.1}$$

Percentage Transmittance, $\%T$ $\qquad \%T_\lambda = T_\lambda \times 100 \tag{5.2}$

Absorbance (for the wavelength λ) is defined by the following:

$$A_\lambda = -\log(T_\lambda) \tag{5.3}$$

In order to calculate T_λ or A_λ, the instrument will need to record both $I_T(\lambda)$ and $I_R(\lambda)$.

In analytical measurements, the value of the absorbance is often a more useful quantity than the transmittance, because the former is more directly related to the concentration of the analyte in the sample — *Beer's Law*, as follows:

$$A_\lambda = a_\lambda bc \tag{5.4}$$

where a_λ is the absorptivity of the analyte at wavelength λ, b is the pathlength through the sample, and c is the concentration of the analyte.

The value of absorbance, A, is often 'calculated' from 'T' by using one of a number of different techniques, as follows:

 (i) a non-linear scale on the display;
 (ii) a logarithmic amplifier in the electronics which converts from T to A;
(iii) the computational power of a microprocessor.

DQ 5.2

A scanning instrument measures $I_T(\lambda)$ and $I_R(\lambda)$ for several wave-lengths, and calculates T_λ and A_λ. Some of these values are given in the table below, while some are missing. Calculate the missing values from the rest of the data.

Wavelength, λ (nm)	λ_1	λ_2	λ_3	λ_4	λ_5
$I_T(\lambda)$ (mW)	0.18		0.31	0.42	
$I_R(\lambda)$ (mW)	0.9	2.5			5.0
$T_\lambda(\%)$		1.5	8.0		
A_λ				1.3	3.2

Answer

The missing values (in random order) are as follows: 0.038 mW; 0.70; 3.88 mW; 0.06%; 1.10; 20%; 0.003 mW; 1.82; 8.4 mW; 5%. NB — the units should help you to identify which is which!

It is important to remember, when evaluating uncertainties in performance, that a spectrophotometer measures *transmittance* directly, and *not* absorbance.

A *single-beam system*, as shown in Figure 5.1, requires that the two samples (test and reference) must be physically *interchanged* so that the two signals, $I_T(\lambda)$ and $I_R(\lambda)$ can be recorded. A *double-beam* spectrophotometer, however, has two separate beams of light, i.e. one for the reference sample and one for the test sample, thus allowing the samples to remain in place during the whole measurement process. The operation of the double-beam system is discussed below in Section 5.3.

5.2 Operation of a Single-Beam Instrument

Is it necessary to interchange the reference sample and test sample every time the wavelength is changed?

It is necessary to measure both $I_T(\lambda)$ and $I_R(\lambda)$ separately to obtain T_λ. The sequence of steps in the measurement procedure depends on whether the instrument has a built-in microprocessor 'memory'.

5.2.1 Without Microprocessor Memory

In this case, the sequence of steps is as follows (see Figure 5.3):

(1) Ensure that, when the light is completely blocked (i.e. when there is no light transmitted), the instrument reads 0% transmittance T. If it does not,

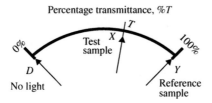

Figure 5.3 Measurement of transmittance for a single-beam instrument without micro-processor control.

then it is necessary to adjust the $0\%T$ (or ∞ absorbance (A)) setting before proceeding.

(2) Set the chosen wavelength, λ.

(3) With the reference sample in place, the light transmitted is $I_R(\lambda)$, and the sensitivity control is then adjusted so that the output display reads $100\%T$ (or $0A$).

(4) The reference sample is replaced by the test sample. The light transmitted is now $I_T(\lambda)$, and the detector output will drop in proportion to the ratio $I_T(\lambda)/I_R(\lambda)$, and the display will give a direct reading between $0\%T$ and $100\%T$ for the unknown constituent (∞A to $0A$).

The setting for $0\%T$ confirms that the amplifier shows $0\%T$ when zero light falls on the detector. This should not change significantly during routine use. It is only necessary to check occasionally that this 'zero offset' of the amplifier is not drifting.

In the above sequence for a non-scanning instrument, it is necessary for the operator to reset the instrument to $100\%T$ (or $0A$) for $I_R(\lambda)$ whenever the *wavelength is changed*. However, this is time consuming for the operator.

DQ 5.3

In a single-beam spectrophotometer set at 500 nm, the sensitivity control is adjusted to give an output reading of $100\%T$. The wavelength setting is then changed to 550 nm, and it is observed, **before adjusting** the sensitivity control again, that the output reading has changed. Which of the following readings is most likely to be the observed value on the $\%T$ scale:

(i) 125%;
(ii) 100%;
(iii) 75%?

Hint — look at the spectral output from a thermal emitter shown later in Figure 17.1.

Answer

It can be seen from Figure 17.1 that by moving from 500 to 550 nm, the output, $I_0(\lambda)$, from the source increases, and this will increase the signal at the detector. As the reading on the transmittance scale is proportional to signal intensity, we can see that the most likely reading out of the options given will be 125%.

5.2.2 With Microprocessor Memory

We see above that $I_R(\lambda)$ must be measured using the reference sample for every wavelength used in the measurement. If the spectrophotometer has a microprocessor with a *memory*, it will be possible for the instrument to record and store the value(s) of $I_R(\lambda)$ at a range of particular wavelength(s) before making any measurements of $I_T(\lambda)$. This process is called 'baseline storage'.

When the test sample is introduced, the instrument records the value of $I_T(\lambda)$, and, using the *stored* value of $I_R(\lambda)$, calculates the result for $\%T$ and A. This process does not require operator adjustment of the sensitivity controls.

DQ 5.4

If, in a single-beam instrument, the intensity of the source gradually decreased with time (drift), what effect would this have on the apparent reading for the transmittance and absorbance of the test sample?

Answer

If the source intensity drops after the system has been set to 100%T using the reference sample (step (3) above), then it will appear that the sample absorbs more light than it actually does. The apparent transmittance will be less than the true transmittance and the apparent absorbance will increase.

It is clear from the above discussion question that the accuracy of the baseline-storage method requires that the source intensity does not drift significantly.

5.3 Double-Beam Systems

A *single-beam* system has the inherent disadvantage that there is a significant time lag between measurements of $I_R(\lambda)$ and $I_T(\lambda)$. Any drift in source output, or detector sensitivity, will give a direct error (DQ 5.4) in the measurement of T, and hence A.

A *double-beam* system is designed to compensate for the effect of source drift. The basic layout is similar to that of the single-beam instrument except that, in the sample compartment area, the single-beam is divided into two beams.

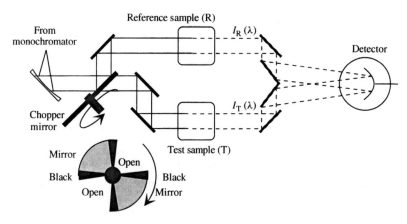

Figure 5.4 Schematic representation of a double-beam UV–visible spectrophotometric system.

Figure 5.4 illustrates the basic design for such a system, although particular instruments may vary in detail. The reference sample, R, and the test sample, T, are positioned separately, one in each of the two beams of light. The two beams are generated alternatively from the same incoming radiation by using the rotating 'chopper' mirror.

A typical chopper mirror may consist of six segments, two of which reflect the beam through the reference sample, two are cut away to allow the beam to pass through the test sample, and two are absorbing segments which prevent any radiation from reaching the detector.

The intensities, $I_R(\lambda)$ and $I_T(\lambda)$, of the two beams from the reference and test samples, respectively, are alternately recorded by the same detector, with a brief period of total absorption (darkness) between each beam. The form of the output current, i, from the detector, as a function of time, is shown in Figure 5.5.

A 'dark current', i_D, (see Section 18.1) occurs when both beams are blocked. The intensities of the two beams can then be derived from the signal, as follows:

$$I_T(\lambda) \propto i_T - i_D \qquad (5.5a)$$

$$I_R(\lambda) \propto i_R - i_D \qquad (5.5b)$$

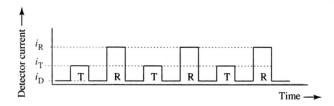

Figure 5.5 Detector output current as a function of time for a double-beam instrument.

The ratio, $I_T(\lambda)/I_R(\lambda) = (i_T - i_D)/((i_R - i_D)$, is 'calculated' by the electronics to give the transmittance, T, of the test sample. This process is known as *ratio recording*.

DQ 5.5

How does the double-beam system compensate for source drift?

(a) It removes the need to keep interchanging the reference and test samples, as is the case with a single-beam instrument.

(b) It is possible to average the signals over a period of time.

(c) The time between making reference and test measurements is reduced to a fraction of a second.

Answer

*The key to the problem of 'drift' is to reduce the **time** between reference and test measurements. The double-beam facility makes this time difference a fraction of a second (c), instead of, perhaps, several minutes. It is an advantage not to keep swapping the samples (a), but this is not the main reason for using the double-beam system. The second answer (b) has no relevance to this particular question.*

The double-beam system is re-calibrating itself every time it switches between the two beams, and the effects of drift ($1/f$ noise) (see Section 14.3) are dramatically reduced. Double-beam instruments are therefore capable of given higher photometric accuracy than single-beam instruments.

5.4 Wavelength Scanning

It is impracticable to plot out a detailed absorption spectrum by using a *manual single-beam* instrument, because it would be necessary to swap reference and test samples at *every* wavelength. However, the *double-beam* system automatically makes a measurement within fractions of a second, as it switches between $I_R(\lambda)$ and $I_T(\lambda)$. With the double-beam system, it is possible to gradually change the wavelength to produce a continuous plot of transmittance (or absorbance) against wavelength.

Before the development of microprocessor memories for single-beam scanning, the double-beam system was the only system that allowed effective wavelength scanning. However, many *single-beam* instruments are now capable of performing wavelength scanning by using *baseline storage* (Section 5.2.2). Single-beam scanning instruments have become more viable because of increases in the overall stability of the system components. Nevertheless, double-beam systems continue to be inherently more accurate for scanning purposes because of the continual comparison between $I_R(\lambda)$ and $I_T(\lambda)$ as the instrument records the two separate beams.

5.5 System Performance

What should I test before I can be confident that my system is working correctly?

A UV–visible spectrophotometer is required to record:

(i) the *absorbance* of an analyte;
(ii) at a specific *wavelength*, or as a function of wavelength;
(iii) by using a *narrow* band of wavelengths for each measurement.

The tests for instrument performance must therefore confirm that the *experimental conditions* (Section 3.1.1), and instrument *response* (Section (3.1.2) are sufficiently accurate and precise. Table 5.1 identifies the *direct* dependencies between the *method characteristics* (Section 1.5) and the *instrument performance characteristics*. Remember also that uncertainties in the method can also be caused by interactions (Section 3.4) between different characteristics.

Table 5.1 Method and system characteristics for a spectrophotometer system

Method characteristics	Primary system characteristics
Accuracy and precision	Wavelength accuracy[a]
	Photometric response
	photometric accuracy and precision
	noise
	stray light
	baseline stability and flatness
Range	Wavelength range[a]
	Photometric range
Linearity	Photometric linearity
Detection limit	(As for photometric response)
Selectivity	Spectral bandwidth[a]
Robustness	Mainly dependent on other aspects of the method rather than the instrument itself

[a]Experimental conditions.

The aim of a System Suitability Test (Section 4.11) is to confirm that the *instrument* performance will satisfy the requirements of the *analytical* method. The most common tests for spectrophotometers usually include some, or all, of the following set of (top-down) tests:

(i) Experimental Conditions — wavelength accuracy and precision, spectral bandwidth, and stray light.

(ii) Instrument Response — photometric accuracy, precision and linearity, photometric noise, and baseline stability and flatness.

The specific tests and the instrumental factors that contribute to these performance
characteristics are presented in the following sections.

For the purposes of Equipment Qualification (Section 4.12), the spectrophotometer is regarded as being a single instrumental unit. The range of performance
tests required for Operational Qualification and Performance Qualification in a
UV–visible spectrophotometer will be similar to those also applicable to System
Suitability Testing. Contrast this with the modular construction of HPLC and GC
systems described in Section 8.7, where it is possible to have different tests at
system and module levels.

5.6 Spectral Characteristics

5.6.1 Wavelength Accuracy and Precision

Wavelength accuracy and precision can be measured by using standard calibrated
spectra (e.g. employing a holmium filter) that give well-defined wavelength peaks
and troughs (see Section 13.7).

Wavelength accuracy depends on the quality of the mechanical and optical
engineering employed in the design and construction of the *monochromator* (see
Chapter 16). In particular, the need to *rotate* the *diffraction grating* to select
the desired wavelength leads to imprecision through friction and backlash. This
requirement for engineering precision and quality is a significant contribution to
the cost of an accurate spectrophotometer.

The problem of moving parts in the monochromator is well illustrated by the
advantage enjoyed by array detector systems, e.g. diode-array (Section 9.3.2)
and photodiode-array (Section 18.3.1) detectors and charge-transfer devices
(Section 18.4), which do not require *any moving parts* in the optical system.
These have excellent wavelength accuracy, precision and long-term reliability.

5.6.2 Spectral Bandwidth

The spectral bandwidth of the spectrophotometer determines its ability to resolve
two close spectral lines. There are two possible consequences (see Section 15.5.1)
of a spectral bandwidth that is *not narrow* in comparison to the width of a spectral
line, as follows:

• broadening of the observed line;
• reduction in the observed absorbance height of the peak.

As a rule of thumb, the spectral bandwidth should not be more than 10% of the
linewidth being measured.

Spectral bandwidth is determined by the physical slitwidth of the monochromator (see Section 16.2.3). This may be fixed, or adjustable within certain limits. The standard test involves the measurement of the ratio between the peak and trough absorbances at 266 and 269 nm in 0.020 vol% toluene in hexane. As the spectral bandwidth increases, these are less well resolved and the ratio between the absorbances at the two wavelengths falls.

DQ 5.6

Spectral bandwidth can have a significant effect on the experimental conditions (see Section 3.1.1) of absorbance measurement, and its value needs to be confirmed. Why then is it not necessary to check the spectral bandwidth as often as the wavelength accuracy?

Answer

A significant change in instrument bandwidth would normally require a mechanical failure (e.g. the knife edges that determine the slitwidth might become loose) which is unlikely to be caused by simple mechanical wear. Hence, the value of the spectral bandwidth does not need to be checked frequently unless a problem is suspected. However, inaccuracies in wavelength setting can be caused by simple day-to-day wear and so the latter must be checked regularly.

5.6.3 Stray Light

Where does stray light come from — inside or outside of the instrument?

Stray light is radiation, arriving at the detector, that has either (i) not passed through the sample, or (ii) is of a different *wavelength* to that being used.

Modern spectrophotometers can be designed to ensure that virtually all of the light reaching the detector has passed through the sample and the monochromator. In practice, therefore, stray light is mainly light of the *wrong wavelength* emerging from the monochromator (see Section 16.2.5), and passing through the sample.

The stray light figure, S_L, is given by the following equation:

$$S_L = \frac{\text{Intensity of } Unwanted \text{ radiation}}{\text{Intenstity of } Wanted \text{ radiation}} = \frac{I_S}{I_0(\lambda)} \qquad (5.6)$$

where I_S is the intensity of the stray light.

The 'wanted' radiation will be the output from the source at the test wavelength. However, because the output from the source, $I_0(\lambda)$, is wavelength-dependent, the stray light figure, S_L, also depends on the wavelength, λ_S, at which the test is carried out. The values quoted for stray light in the specifications of instruments are given for measurements at particular wavelengths, e.g. by using *NaI (10 g/l) at 220 nm*.

The ideal *test sample* for measuring stray light at λ_S would be a material with the following characteristics:

- a very high absorbance over a narrow wavelength range (around λ_S);
- a very low absorbance at all other wavelengths.

The stray light can then be calculated from the *apparent* absorbance, $A_S' = -\log S_L$.

Stray light will add to both $I_T(\lambda)$ and $I_R(\lambda)(\approx I_0(\lambda))$, and increases the apparent transmission, T_{app}, of the sample (Section 5.1) as follows:

$$\text{Apparent Transmittance, } T_{app}(\lambda) = \frac{I + I_S}{I_0 + I_S} \tag{5.7}$$

where $I = I_T(\lambda)$, and $I_0 = I_R(\lambda) \approx I_0(\lambda)$.

The possible uncertainty in transmittance is given by the following equation:

$$u'(T) = T_{app} - T = \frac{I + I_S}{I_0 + I_S} - T = \frac{I/I_0 + I_S/I_0}{1 + I_S/I_0} - T = \frac{T + S_L - T(1 + S_L)}{1 + S_L} \tag{5.8}$$

Since S_L is usually very small, we can put $(1 + S_L) \approx 1$, and can then express the uncertainty in T, due to stray light, in the following way:

$$u'(T) \approx [S_L - (T \times S_L)] = S_L(1 - T) \tag{5.9}$$

Note that $u'(T)$ is *not* a *random* uncertainty (Section 2.4), and its magnitude will depend on the wavelength being used for the measurement.

The relationship between true and apparent absorbance is shown in Figure 5.6 for a value of stray light, S_L, of 0.001. Note that, for a stray light figure of S_L, it is impossible for the instrument to record an apparent absorbance higher than $A_S = -\log S_L$.

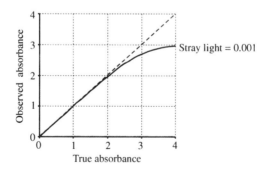

Figure 5.6 The relationship between true and observed absorbance, showing the effects of stray light; $S_L = 0.001$.

DQ 5.7

An ideal material for measuring stray light at wavelength λ_S, has a virtually infinite absorbance at this wavelength, and a very low absorbance at all other wavelengths. If the spectrophotometer records an apparent absorbance of $3.0A$ at this wavelength, what is the equivalent stray light figure, S_L:

(a) 0.33;
(b) 0.01;
(c) 0.033;
(d) 0.001?

Answer

The apparent transmittance, $T_{app} = 10^{-A} = 10^{-3} = 0.001$. In addition, $T_{app} = I_S/I_0 = S_L$, and hence $S_L = 0.001 = 0.1\%$.

SAQ 5.1

A spectrophotometer is specified as having a stray light figure of $< 0.05\%$. Calculate the possible error in absorbance for a particular sample that shows an apparent absorbance of $2.2A$.

Would the *percentage* error be the same at other absorbance values?

5.7 Photometric Uncertainties in Absorbance and Transmittance

What effect does photometric uncertainty in the instrument have on the measurement of concentration?

We have seen above (Section 5.1) that the spectrophotometer actually measures the *transmittance*, T. Any uncertainties in instrument performance will appear as uncertainties in transmittance, i.e. $u(T)$.

However, the analyst is primarily concerned with the uncertainty in the *final results*. As the final result is often a measurement of concentration, c, then the important figure of merit is as follows:

Relative Uncertainty in Concentration

$$= \frac{\text{Uncertainty in the value of concentration, } u(c)}{\text{Value of concentration, } c} = \frac{u(c)}{c} \quad (5.10)$$

Assuming Beer's Law, the concentration, c, is proportional to the absorbance, A, and the *fractional* uncertainties (Section 2.4.3) in c and A will be the same, as follows:

$$u(c)/c = u(A)/A \tag{5.11}$$

Note that if $u(A)$ represents the noise in the absorbance signal, then $u(A)/A$ is the *reciprocal* of the signal-to-noise ratio (S/N).

If we now wish to quantify this relative uncertainty, we must find out how $u(A)/A$ is related to $u(T)$. We know from equation (2.13) that the following relationship applies:

$$u(A) = \frac{dA}{dT} \times u(T) \tag{5.12}$$

and the relationship between A and T is given by the following:

$$A = -\log T = -\ln T/2.30 \tag{5.13}$$

Differentiating this equation, we obtain the following:

$$dA/dT = -1/(2.30 \times T) \tag{5.14}$$

Using the uncertainties, $u(A)$ and $u(T)$, expressed only as positive values, and dividing by A, we then obtain the following:

$$\frac{u(A)}{A} = (-)\frac{u(T)}{2.30 \times T \times A} \tag{5.15}$$

Note that equation (5.15) applies to *small uncertainties*, and assumes that $u(A) \ll A$.

The positive values are used because the sense (direction) of random uncertainty is unpredictable. The negative sign will also disappear when 'u' is squared to combine the effect randomly with other uncertainties (Section 2.4). The negative relationship between $u(A)$ and $u(T)$ is retained (as a negative sign in parentheses) for use in situations where a non-random association needs to be calculated.

If we derive equations for $u(T)$, as in the sections below, we can calculate the possible fractional uncertainties in A and hence the fractional uncertainties in c.

DQ 5.8

A spectrophotometer has a 1% uncertainty in measuring T. Which experimental condition (i.e. approximate value of A) gives the greatest fractional uncertainty in the absorbance A:

(a) $A = 0.1$;
(b) $A = 1.0$;
(c) $A = 3.0$?

Answer

A 1% error in T gives u(T)/T = 0.01. By substituting this into equation (5.15), it is possible to calculate the fractional uncertainty in

A, u(A)/A, *for each value of* A, *to give (a) 0.0435, (b) 0.00434 and (c) 0.00145. The greatest fractional uncertainty in this case occurs with the smallest value of absorbance, i.e. with only a very small concentration of analyte (see also the effect of source drift (Section 5.72 and Figure 5.7).*

5.7.1 Tracing Uncertainties in Absorbance and Transmittance

Referring to the analogue scale shown above in Figure 5.3, the measurement of transmittance, T, requires the recording of signals equivalent to the following:

(i) D — the reading when the light is completely blocked;
(ii) Y — the reading with the Reference Sample in place;
(iii) X — the reading with the Test Sample in place.

The value equivalent to $I_R(\lambda)$ for the Reference Signal is $(R = Y - D)$. This has the following uncertainty:

$$u(R)^2 = u(Y)^2 + u(D)^2 \qquad (5.16)$$

The value equivalent to $I(\lambda)$ for the Test Signal is $(S = X - D)$, which has an uncertainty as follows:

$$u(S)^2 = u(X)^2 + u(D)^2 \qquad (5.17)$$

The calculated value of the transmittance T is given by S/R. Hence, the estimate for the uncertainty, $u(T)$, can be calculated (Section 2.4.2) by using the following expression:

$$\left[\frac{u(T)}{T}\right]^2 = \left[\frac{u(S)}{S}\right]^2 + \left[\frac{u(R)}{R}\right]^2 = \frac{1}{R^2}\left[u(S)^2\left(\frac{R}{S}\right)^2 + u(R)^2\right] \qquad (5.18)$$

Substituting for $T(= S/R)$, $u(S)$ and $u(R)$, and then rearranging gives the following:

$$u(T) = \frac{1}{R}\sqrt{\left\{u(X)^2 + u(D)^2 + \left[u(Y)^2 + u(D)^2\right]T^2\right\}} \qquad (5.19)$$

The above equations permit a calculation of the combined effect of uncertainties in the three signals that contribute to the value of the transmittance T. This derivation assumes that the uncertainties in each signal are *random* and *independent* of each other.

5.7.2 Drift in Source Output

The performance of all components changes slowly with time — a process which is known as 'drift' (see Section 14.1). In particular, the intensity of radiation from the *source* will tend to change, thus giving a drift in the *zero-absorbance* setting.

If after measuring Y for the Reference Sample, the source intensity *decreases* before measuring X for the Test Sample, it will *appear* that the Test Sample has transmitted less light than is really the case. The calculation of apparent transmittance will give a lower value than the true transmittance. If, however, the output from the source increases, then the apparent transmittance, T_{app}, will be greater than the true value.

Changes in the *response of the detector* and the *gain of the amplifier* will also have an effect similar to a change in source output. However, these components tend to be more stable than the source and are not normally the main cause of error.

Referring again to Figure 5.3, we can make the following statements concerning the drift in the *source* output:

- it can not affect D because all of the light is blocked, i.e. $u(D) = 0$;
- it does not affect Y because the process of measurement (Section 5.3) sets the instrument to read 100%T (0A) when recording Y, i.e. $u(Y) = 0$;
- it gives an uncertainty $u(X)$ when the Test-Sample signal, X, is measured.

If the source intensity increases by a *factor*, k_D, then the uncertainty in X is given by the following:

$$u(X) = k_D X \tag{5.20}$$

Substituting into equation (5.19) we find the following:

$$u_D(T) = \frac{1}{R}\sqrt{[u(X)^2]} = \frac{u(X)}{R} = \frac{k_D X}{R} \approx k_D T \tag{5.21}$$

We can make the final approximation in the above equation because $X/R \approx T$. When this relationship is substituted into equation (5.15), we then obtain the following relationship:

$$\frac{u(A)}{A} = (-)\frac{k_D T}{2.30 \times T \times A} = (-)\frac{k_D}{2.30 \times A} \tag{5.22}$$

Note that the negative sign in parentheses shows that an *increase* in source intensity (k_D positive) gives an apparent value of absorbance that is *lower* than the true value.

5.7.3 Thermal (Johnson) Noise in the Detection System

The origins of thermal noise are described later in Section 14.4.1. In a spectrophotometer, thermal noise will be most significant for radiation of low photon energy, i.e. it will be important in IR and less important in UV–visible spectrophometric systems.

The magnitude of thermal noise in a detector does not depend on the strength of the signal arriving at the detector. Thus, the uncertainty, k_J', due to random thermal noise is independent of the *magnitude of the signal at the detector*. It will be the same for X, Y and D, and so we can write the following:

$$u(X) = u(Y) = u(D) = k_J' \qquad (5.23)$$

By substituting into equation (5.19), we then obtain the following:

$$u_J(T) = \frac{1}{R}\sqrt{[2(k_J')^2 + 2(k_J')^2 T^2]} = \frac{\sqrt{2}k_J'}{R}\sqrt{(1 + T^2)} = k_J\sqrt{(1 + T^2)} \qquad (5.24)$$

where:

$$k_J = \sqrt{2}k_J'/R \qquad (5.25)$$

Note that the *effect*, k_J, of thermal noise is *reduced* if the signal strength, $R(= I_0(\lambda))$, can be *increased*. When this relationship is substituted into equation (5.15) we obtain the following:

$$\frac{u_J(A)}{A} = \frac{k_J\sqrt{(1 + T^2)}}{2.30 \times T \times A} \qquad (5.26)$$

The negative sign in parentheses has been dropped because thermal noise is random in character, and thus it is not possible to predict the direction of the effect on 'A'.

5.7.4 Shot Noise in the Detection System

Shot noise is significant in the 'photon' detectors used in UV–visible spectrophotometers. We shall see later in Section 14.4.2 that the magnitude of shot noise is proportional to the square root of the current in the detector, where this current is proportional to the magnitudes, Y, X and D, of each of the signals. Hence we can write the following:

$$u(Y) = k_S'\sqrt{Y}; u(X) = k_S'\sqrt{X}; u(D) = k_S'\sqrt{D} \qquad (5.27)$$

Substituting into equation (5.19), we can derive the following expression:

$$u_S(T) = \frac{k_S'}{\sqrt{R}}\sqrt{\left[\frac{(X + D)}{R} + \frac{(Y + D)}{R}T^2\right]} \qquad (5.28)$$

Since $S = X - D$, we know that $X + D = S + 2D$, and similarly that $Y + D = R + 2D$. Hence, by substitution and rearranging we obtain the following:

$$u_S(T) = \frac{k_S'}{\sqrt{R}}\sqrt{\left[T + T^2 + (1 + T^2)\frac{2D}{R}\right]} \qquad (5.29)$$

Provided that the minimum signal measurement, $S(= X - D)$, is significantly greater than the dark current signal, D, we can then write the following:

$$\frac{S}{R} \gg \frac{2D}{R} \qquad (5.30)$$

Hence $T \gg D/R$, and introducing $k_S = k_S'/\sqrt{R}$, we can make the following approximation:

$$u(T)_S \approx k_S \sqrt{(T + T^2)} \qquad (5.31)$$

Note that the *effect*, k_S, of shot noise is reduced if the *signal strength*, $R(= I_0(\lambda))$, can be *increased*. When this relationship is substituted into equation (5.15), we obtain the following:

$$\frac{u_S(A)}{A} = \frac{k_S \sqrt{[T + T^2]}}{2.30 \times T \times A} \qquad (5.32)$$

The negative sign in parentheses has been dropped because shot noise is random in character, and so it is not possible to predict the direction of the effect on 'A'.

DQ 5.9

Which of the following is (are) generally true for a spectrophotometer:

(a) decreasing the slitwidth will increase noise;
(b) noise will increase with the age of the source;
(c) noise will depend on the wavelength being used?

Answer

*The relative effects of thermal noise, k_J, and shot noise, k_S, both increase if the signal strength **decreases**. Signal strength will decrease as the square of any decrease in slitwidth (see Section 16.2.4), thus resulting in a decreased signal-to-noise ratio. As the source ages, it does not itself become necessarily noisier, but the resulting drop in power output will increase the relative noise in the detector system. Similarly, the power output from the source will also be a function of wavelength, e.g. the drop in power output at low wavelengths prevents the use of tungsten lamps below 320 nm. Hence, all of the factors are likely to affect the signal strength, and hence the apparent level of noise in the system.*

5.7.5 Combined Photometric Uncertainties

The behaviour of the two types of noise is shown in Figure 5.7, which plots the fractional uncertainty, $u(A)/A$, as a function of A for the different types of noise, as well as drift. Arbitrary values have been chosen for the 'k' constants in order to show the relative effects of the different types of uncertainty on the same graph scale.

The effect of stray light (Section 5.6.3) is also included for comparison, i.e. $u_L'(T) = S_L(1 - T)$, which gives the following:

$$\frac{u_L'(A)}{A} = (-)\frac{S_L(1 - T)}{2.30 \times T \times A} \qquad (5.33)$$

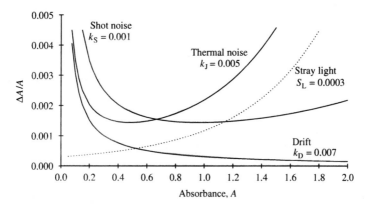

Figure 5.7 Fractional photometric uncertainty as a function of absorbance for thermal and shot noise, and drift in source output; the behaviour of stray light is also shown for comparison.

However, it must be remembered that stray light is not a *random* uncertainty and will always give a lower apparent absorbance than the true value. The uncertainties due to noise and drift are random.

It can be seen from Figure 5.7 that the fractional uncertainty in A is not the same for all values of the absorbance. A low value of $u(A)/A$ gives the least relative uncertainty in the value of concentration. Hence, there is a considerable advantage in planning the analysis (e.g. any dilution procedure) such that the absorbance to be measured corresponds to (or is near to) the value of A that gives the minimum fractional uncertainty.

DQ 5.10

What is the optimum value of absorbance at which to make a photometric measurement when using a UV–visible spectrophotometer with a photomultiplier detector:

(a) less than $0.3A$;
(b) about $0.45A$;
(c) about $1.0A$;
(d) greater than $2.0A$?

Answer

The limiting noise in a photomultiplier system is shot noise. It can be seen from Figure 5.7 that, providing other noises can be reduced sufficiently, shot noise gives the best signal-to-noise ratio at about $1.0A$. If, however, the limitation was thermal noise, then the optimum absorbance value would be about $0.45A$.

SAQ 5.2

Calculate the dynamic range for measurements made by using a spectrophotometer which is limited by source drift with $k_D = 0.007$ and stray light with
$S_L = 0.0003$ (see Figure 5.7), assuming that the minimum acceptable uncertainty
is 0.4%.

5.8 Photometric Characteristics

5.8.1 Accuracy, Precision and Noise

Figure 5.7 shows the behaviour of the fractional uncertainty, $u(A)/A$, due to
noise and stray light, as a function of *absorbance*, *A*. However, as the source and
detector characteristics are also *wavelength-dependent*, the fractional uncertainty,
$u(A)/A$ will also vary with *wavelength*. Thus, a *full specification* of photometric
uncertainty would require the values to be plotted on a two-dimensional 'map'
covering the full ranges of both absorbance and wavelength.

In practice, the values for photometric *noise* are quoted for specific values of
both *absorbance and wavelength*. However, it is important to remember that the
photometric uncertainty will normally be greater at wavelengths towards the end
of the operational wavelength range and at large values of absorbance.

Photometric *accuracy* is normally quoted at an absorbance value of 1.0*A*,
possibly with the performance at other values of absorbance also being included,
e.g. as follows:

*Photometric accuracy at 0.5*A: ±0.003*A*;
*Photometric accuracy at 1.0*A: ±0.004*A*;
*Photometric accuracy at 2.0*A: ±0.012*A*.

DQ 5.11

Using the example specifications given above, at what value of the
absorbance, *A*, would the spectrophotometer give the most accurate
measurement of concentration?

Answer

*The important factor is the **fractional** uncertainty in absorbance, u(A)/A.
Calculating these ratios, we find that the values are 0.006, 0.004 and
0.006, respectively. The lowest value of dA/A (i.e. the most accurate)
occurs for A = 1.0.*

Some manufacturers simulate the variation of photometric accuracy with
absorbance by using an *expression* to give the limiting specification for
uncertainty, e.g. as follows:

Photometric accuracy — 0.5% or 0.003A, whichever is greater up to 3.0A.

SAQ 5.3

Identify which of the following values for noise, i.e. $< 0.0002A$, $< 0.0004A$, $< 0.0008A$ or $< 0.002A$, might apply to the following conditions in a good quality spectrophotometer:

(a) noise at 0A with 2 nm bandwidth;
(b) noise at 0A with 5 nm bandwidth;
(c) noise at 2A with 2 nm bandwidth;
(d) noise at 2A with 5 nm bandwidth.

Can you suggest which is which?

The response 'characteristics' can be measured by using standard samples that give well-defined absorbance levels. The absorbance standard can be a calibrated glass filter (e.g. Schott NG at 546 nm) or a standard solution ($K_2Cr_2O_7$ at specific peak wavelengths). These have spectra with absorbance values that change very slowly with wavelength.

5.8.2 Baseline Flatness and Baseline Stability (Drift)

The 'baseline' in an absorbance spectrum represents the measurement of zero absorbance or 100% transmittance. The displayed value of 0A for zero absorbance should be '0' for all values of wavelength, and should remain at '0' over a long period of time.

Baseline Stability (or *Drift*) gives the maximum change (drift) in '0A' with *time*, e.g. $\pm 0.0001A$/h, and can be measured simply by observing the gradual change in a near-zero absorbance reading. For a single-beam instrument, the baseline drift is due mainly to a drift in the output from the light source.

Baseline Flatness measures the maximum error in '0A' with *wavelength*. It may be expressed as the uncertainty limit over the whole wavelength range, e.g. $\pm 0.003A$.

A double-beam system compensates for source drift (Section 5.3) and provides very good baseline *stability*. However, although an exact compensation may be achieved at a *specific* wavelength, the two optical paths may be affected differently by a change in wavelength, e.g. by different reflectances in the two beams. The consequence of this imbalance would be a slight shift away from '0A' for different wavelengths, i.e. a non-zero value for baseline *flatness*.

5.8.3 Photometric Linearity

The linear response range (Section 3.2.4) is normally limited by both the lower limit (the detection limit) and the upper limit of the instrument response. Figure 5.7 shows how uncertainties at both high and low values of the absorbance set limits to the useful range of the spectrophotometer. However, the lower detection limit for spectrophotometric *methods* varies considerably across the wide range of possible analytes, and it is not appropriate to define it in terms of a particular analyte. Hence, it is not useful to define linearity as a dynamic range between upper and lower values.

Therefore the linearity range is usually described by defining the *upper limit* for absorbance for which the responsivity of the system remains within a defined range, e.g. *linearity* > 3A.

5.8.4 Operational Wavelength Range

The operational wavelength range is defined by the manufacturer as that range of wavelengths over which the instrument is capable of producing an acceptable photometric accuracy.

Good photometric accuracy is dependent on sufficient signal strength to reduce the relative effects (Section 5.7) of detector noise and stray light. Thus, the range of wavelengths over which a UV–visible instrument can operate is limited mainly by the performances of its sources and detectors, and the transmittance of its optical components. The photometric accuracy will be less (i.e. worse) near the ends of the wavelength range. If the overall source output or detector responsivity were to fall (possibly due to age), then significant photometric errors would first appear at the ends of the spectral range.

The short-wavelength end (350 nm) of the *visible* range is limited by the output from the tungsten source (see Section 17.2). The performances of these instruments fall quickly at very short wavelengths due to the rapid drop in source energy. At the long-wavelength end, the spectrophotometer operates into the near-IR region, with a limit which is normally given by the response range of the detector (phototube and photomultiplier up to about 900 nm, or photodiode up to about 1100 nm).

The addition of a deuterium source (see Section 17.3.1), which can be switched in as required, allows the wavelength range to be extended into the UV region. Pulsed xenon sources (see Section 17.3.2) have now been developed to operate across the whole UV–visible range.

SAQ 5.4

Identify the possible options for the basic design of a UV–visible spectrophotometer which will be capable of *wavelength scanning*.

SAQ 5.5

How may a knowledge of the spectrophotometer photometric performance impact on the design of the analytical *method*?

Note. Further questions that relate to the topics covered in this chapter can be found in SAQs 2.2, 2.5, 3.4, 7.3, 8.4, 11.4, 15.3, 16.3, 17.3 and 18.2, and DQs 1.3, 2.5, 3.1, 3.9, 3.10, 4.3, 9.7, 14.3, 14.6, 15.4 and 15.5.

Summary

This chapter has discussed the main spectral and photometric *specifications* appropriate to selecting a UV–visible spectrophotometer. The relationship between the analytical *method* characteristics and the *instrument* characteristics was introduced, and the approach to Equipment Qualification for a spectrophotometer was contrasted to that used for an HPLC instrument.

The use of microprocessor memories in single-beam instruments now provides an alternative method of wavelength scanning without the need for a double-beam system, and hence special attention was paid to the advantages and disadvantages of *single-beam* and *double-beam* systems.

An additional core theme to this chapter was the *variation of uncertainty* with operational settings, and the implied benefits of planning the experimental procedure to match instrument performance. This was addressed quantitatively by examining the effect of noise and stray light on photometric performance.

Chapter 6
Atomic Spectroscopy

Learning Objectives

- To appreciate why different atomization temperatures provide different opportunities in the analysis of elements.
- To identify the main performance features of flame, graphite and plasma systems.
- To differentiate between methods of background correction.
- To identify the main (performance) characteristics required to specify the performance of these instruments.

6.1 Radiation Processes

The practice of atomic spectroscopy, by using optical emission and absorption, depends on the fact that the wavelength of the radiation, λ, emitted, *or* absorbed by an atom is a function of the energy transition, E, occurring within the atom as it moves from one energy state to another, as follows:

$$E = h\nu = hc/\lambda \qquad (6.1)$$

where ν is the frequency of the radiation, c is the speed of light, and h is the Planck constant. The particular energy transitions (and hence wavelengths) are characteristic of the specific atoms.

The radiation emitted or absorbed by a free atom typically has a spectral bandwidth which is less than 0.001 nm. However, this can be broadened by Doppler motion and by pressure broadening (see Section 17.1).

We will see later in Section 17.2 that a good emitter of radiation at a particular wavelength must be a good absorber at the same wavelength. This also applies

on the atomic scale, thus giving the possibility of detecting the atom by its
absorption or *emission* spectrum.

The free atoms to be analysed in atomic spectroscopy are dissociated from their
molecular matrix by a flame, furnace or plasma, with the following approximate
temperatures:

• Flame — air/acetylene at c. 2500 K and N_2O/acetylene at c. 3000 K;
• Graphite furnace — up to c. 3000 K;
• Plasma — 8000 to 12 000 K.

NB — plasma is a 'fourth' state of matter, in which a significant proportion of
the atoms of a *gas* are in an *ionized* state.

At a high temperature, a fraction of the atoms (n/n_0) will each acquire energy,
ΔE, to move from the ground state into a higher energy state. This fraction is
given by an exponential equation which is dependent on the ratio between the
required energy, $\Delta E(= hc/\lambda)$, and the available thermal energy, kT, and is given
by the following equation:

$$\frac{n}{n_0} = \exp(-\Delta E/kT) \tag{6.2}$$

DQ 6.1

Calculate the ratio n/n_0 for both potassium and copper at temperatures of
3000 and 6000 K, given that the wavelengths equivalent to the transition
to the excited states are 766.5 and 324.7 nm, respectively.

Answer

*The first step is to calculate the energies of transition from the equiva-
lent wavelengths, and by using equation (6.1) we find values of $2.59 \times
10^{-19}$ J and 6.10×10^{-19} J, for K and Cu, respectively. Calculating
the values of kT for the two temperatures, we then obtain, respectively,
4.14×10^{-20} J and 8.28×10^{-20} J.*

Hence, by using equation (6.2), the ratios are:

	3000 K	*6000 K*
Potassium	1.9×10^{-3}	4.4×10^{-2};
Copper	4.0×10^{-7}	6.3×10^{-4}.

The above discussion question shows that, at lower temperatures, very few copper
atoms will be in an excited state and thus be capable of *emitting* radiation. This is
true for most elements, and the flame and electrothermal (furnace) sources are not
hot enough to excite a significant proportion of the atoms into an emitting state.
Hence, flame and furnace systems normally operate in the *absorption* mode.

However, for alkali elements the temperature of a *flame* is sufficient to excite enough atoms into high-energy states, and flame *emission* photometry is a well-established technique for quantitative measurement of sodium, potassium and lithium. At higher temperatures, a significant proportion of all elements can move into higher energy states from which they can subsequently become *emitters* of radiation.

Electrical arc discharges were initially used to achieve *emission*. However, these sources are difficult to maintain and calibrate, and have been mainly replaced by *plasma* sources.

Atomic spectroscopy in its various forms is susceptible to a variety of *interferences*, as follows:

- **Chemical interferences,** where the chemical environment in the sample may affect the rate at which atomization occurs and hence the intensity of the signal — a matrix effect.

- **Physical interferences,** where the physical properties of the sample affect the efficiency of atomization.

- **Spectral interferences,** where these include overlapping of spectral emissions and absorptions, due to other species present in the sample, and background emission of the atomizing source (e.g. a flame).

6.2 Absorption Spectroscopy

The basic system, shown in Figure 6.1, for measuring *atomic* absorbance is similar to that of the UV–visible spectrophotometer, but with the sample presented within a flame or furnace.

The process of getting the element, in its atomic form, into the right place includes the following stages:

- *Nebulization,* where the liquid sample is introduced into the gas flow, leading to the flame, as a cloud of fine droplets;

- *Transport to the flame,* where the gas carries the droplets into the flame;

- *Atomization,* where the temperature of the flame needs to be such that the highest possible proportion of the sample is dissociated into the atomic state.

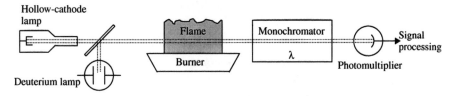

Figure 6.1 Schematic representation of an atomic absorption spectrophotometer.

As the spectral linewidth of the absorption is extremely narrow, it is essential that the spectral bandwidth of radiation used is also very narrow. The ideal source uses narrow-line emission from atoms of the *same element* as the one to be detected. This is achieved with either a hollow-cathode lamp (HCL) or an electrodeless discharge lamp (EDL) (see Section 17.3.3), which both use an electrical discharge in a vapour containing the same specific element(s) to give high-intensity radiation at exactly the correct wavelength.

A disadvantage of both these atomic absorption spectroscopy (AAS) sources is that a separate lamp is usually required for each element. Some lamps act as sources for two or three elements simultaneously, but the radiation intensity from each element is reduced. Although it is possible to have several lamps on a rotating turret, it is still a cumbersome process to measure several elements in the same sample.

The monochromator in an AAS system is not used to define the spectral bandwidth of the measurement, as this is a function of the lamp. Its main purpose is to *reject* the background radiation at wavelengths outside the bandwidth of the lamp.

DQ 6.2

The AAS measurement process has to face a number of problems in experimental design — these are listed as (A), (B) and (C) in the table given below.

In order to overcome some of the problems, an AA spectrophotometer may:

(i) modulate the source and only detect the signal at the modulation frequency;
(ii) use a double-beam system.

Identify in the table which of the problems are addressed by these two techniques and which are not.

		Technique	
Problem	Details	(i)	(ii)
A	The source intensity may drift		
B	Emissions from material in the flame will provide spurious signals		
C	Absorption and scattering by other particles and molecules within the flame will give an increased apparent absorption		

Answer

(i) *Modulation of the source (either by a modulated current or a rotating chopper) ensures that the analytical information is carried by a specific frequency component. Light 'noise' emitted from the flame (or furnace) will contain components at a wide range of frequencies, and most of these will be **rejected** by the tuned detection system (B).*

(ii) *As in UV–visible instruments, the double-beam system compensates for source drift (A).*

However, neither technique (i) or (ii) can distinguish the absorbance of the sample from the background absorbance (C). A more subtle approach is needed, namely background correction (see Section 6.3 below).

6.2.1 Flame Atomization

In the flame system, a nebulizer sprays the sample solution, as a cloud of fine droplets, into the gas flow leading to the flame. Most modern systems use either air/acetylene or N_2O/acetylene as the fuel mixture for the flame, giving temperatures up to about 2500 and 3000 K, respectively. A narrow burner slot (5–10 cm in length) aligns a flat flame with the optical beam.

Atomization is a crucial stage in the overall *responsivity* of the system, as a large proportion (over 80%) of the sample fails to reach the flame. The nebulization process is dependent on various physical properties of the sample (e.g. viscosity, surface tension, etc.). Problems can also occur due to deposition of salts affecting the gas flow through narrow orifices — including the burner slot.

The 'efficiency' of the flame itself, in presenting the element in its atomic form, depends on a number of factors, e.g. flame temperature, relative gas flow rates and height of the burner. These must be optimized *separately* for each particular element.

Flame systems have advantages of low capital costs, short analysis times, high sample throughput, good precision and accuracy, and good detection limits. However, due to interferences, the *absorption* technique tends to have a non-linear response and, consequently, a poor dynamic range. Automatic diluters have been developed to work with the spectrophotometer to provide combined systems that have a better dynamic range.

SAQ 6.1

You wish to measure the concentration of four different elements in several samples by using F-AAS. Which factors influence the order in which you carry out the operation, i.e. do you measure all four elements in each sample before proceeding to the next sample, or do you measure for one element in all samples before proceeding to the next element?

6.2.2 *Graphite Furnace Atomization*

Here, the sample is introduced on to a platform which is positioned inside a small longitudinal furnace. The furnace then follows a controlled temperature programme in order to achieve the following phases:

Phase 1–*Drying*. The liquid sample is heated to the drying temperature (T_D) (about 120°C) for about 20 s. This temperature needs to be high enough to cause rapid evaporation of any liquid, without any violent boiling action that could eject droplets and loose analyte.

Phase 2–*Ashing*. The sample is now heated to the ashing temperature (T_A) (possibly up to 1200°C) in order to remove as much of the matrix material as possible. The control of this stage is very dependent on the particular type of sample, as the aim here is to remove as much non-analyte as possible without removing any of the analyte itself.

Phase 3–*Atomization*. At this stage, the temperature has now reached a maximum in the cycle (up to about 2800°C), leading to the production of a cloud of vaporized atoms which lasts for a few seconds.

Phase 4–*Cleaning*. The temperature is maintained high for a few more seconds in order to remove all remaining sample material from the tube before the next run.

The absorption due to the analyte will peak over a short time period, with the instrument being designed to make an *integrated* measurement over this period.

As the temperature of the sample is raised in the atomization phase, the various components of the sample will be atomized at different times — it is therefore also possible to differentiate between different components (*selectivity*) from the *times* at which the absorptions are measured.

SAQ 6.2

Why does GF-AAS give increased sensitivity over F-AAS?

6.3 Background Correction for Absorption Systems

Background absorption occurs in both flame (F) and graphite furnace (GF)-AAS, due to the *absorption* and *scattering* of light by material other than the analyte being measured.

For GF-AAS in particular, the background absorption is also due to scattering by particles produced by the complete decomposition of the sample. The amount of scattering is dependent on the particular sample matrix, thus making it difficult to produce truly representative blanks.

Background correction is an essential process which allows for this additional *unwanted* absorption. There are three main methods, all of which rely on the fact that absorption by a *free* atom has a very narrow spectral linewidth, i.e. ∼0.005 nm. These are described in the following sections.

6.3.1 Continuous-Source Method

The *average* background absorbance over a wider spectral range (i.e. the spectral bandwidth of the monochromator ≈1 nm) is measured by using a continuous-radiation source (e.g. a deuterium lamp). This value is then used to estimate, and correct for, the effect of the background absorbance over the much narrower spectral bandwidth of the HCL lamp.

In principle, the measurement of background absorbance over the wider spectral range will also include a small fraction due to the analyte absorbance. However, this fraction will only be significant if the analyte signal is very large in comparison to the background absorbance, in which case the error due to the background absorbance will already be low.

This form of background correction generally works very well, except in cases where the background absorption has a very structured profile, i.e. when the background absorbance at the analyte wavelength is not equal to the average absorbance over nearby wavelengths.

6.3.2 Zeeman Method

When a *magnetic field* is applied to an atom, interaction of the electron spin with the field splits the degenerate atomic energy levels into new levels. This produces new transition energies that are slightly different from the initial energy, and the spectral line profile develops 'wings' (σ-components) at wavelengths which are slightly separated from the central (π) component, a process which is known as Zeeman splitting (see Figure 6.2). The radiation waves emitted/absorbed by the σ- and π-components are polarized at angles of 90° relative to one another, and

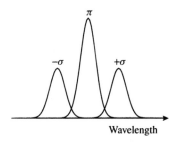

Figure 6.2 Zeeman line splitting.

it is possible, by using polarizer plates, to measure the two radiation components independently.

Although the magnetic field could be applied to either the lamp or the sample, the usual practice is to apply the field to the sample.

With the magnetic field switched on, the *background* absorption can be measured by using a polarizer plate in order to polarize the light from the HCL such that it does not interact with the π-component of the sample. There is then no absorbance from the sample at the central wavelength, and the system records *only* the background absorbance.

The observed absorbance (*analyte* plus *background*) can then be measured by turning off the magnetic field so that the radiation interacts with the σ-component of the analyte. The true analyte absorbance is then obtained by subtracting the known background absorbance.

DQ 6.3

For a GF-AAS system, which factors might determine the rate at which the magnetic field is switched on and off in the Zeeman background-correction method?

Answer

The process of turning the magnetic field on and off takes a finite time, and sampling at a frequency that is too high will reduce the effective time for which the instrument is gathering data. However, the signal obtained from GF-AAS is time-varying, which can not be sampled too slowly or otherwise information will be lost. The sampling frequency must therefore be a compromise, in order to produce the best signal-to-noise (S/N) ratio with the minimum loss of information.

6.3.3 Smith–Hieftje Method

When an HCL is driven with a *high current*, two phenomena can be seen, as follows:

- the emission line becomes considerably broader than the analyte absorption line;
- at the central analyte wavelength, a process of self-absorption by cooler evaporated atoms in the lamp causes a considerable drop in the lamp output.

Thus at high current levels, the HCL becomes effectively a doublet source with wavelengths on *either side* of the analyte absorption. In this mode, it can be used as a source to measure the average *background* absorption on each side of the analyte line.

The analyte absorption, *plus* background, is measured with a normal current through the HCL, with background correction then being achieved by subtracting the background absorption that is measured with the high current. This method was first proposed by Smith and Hieftje [1].

DQ 6.4

What is the significant difference between the way in which the Zeeman method works when compared to the other two methods?

Answer

*Both the continuous-source and the Smith–Hieftje methods work by mea-suring the background on either side of the line, and **interpolating** the value of the background absorbance at the analytical wavelength. The Zeeman effect measures the background absorbance at the **same** wavelength as that of the line itself.*

6.4 Plasma Emission Spectroscopy

Atomic emission for most elements requires a higher temperature than it is possible to achieve with a flame. Electric sparks have been used as the exci-tation source, but a more successful system uses inductively coupled plasma (ICP)-atomic emission spectroscopy (AES) (or ICP-optical emission spectroscopy (OES)).

An ICP 'torch' produces a plasma within a jet of argon gas which is heated by radio frequency (RF) induction from an external RF coil. The very hot plasma is kept in position, away from the walls of the torch, by using an additional outer vortex jet of argon. The sample is introduced into the middle of the plasma along the central axis of the torch. The plasma can be viewed either along the axis of the torch, or perpendicular to this axis.

The outer section of the plasma is cooler than the central section, and it is possible that some re-absorption of the emitted radiation can occur. Two main methods are used to reduce this re-absoption in the axial mode, as follows:

- a shear gas is used to blow away the cooler part of the plasma;
- the plasma is directed on to the apex of a shallow cone which deflects the cooler gas radially, allowing the radiation from the centre of the plasma to pass through a hole in the apex of the cone.

The high temperature of the plasma (8000–12 000 K) is sufficient to excite atoms into many excited states (including ionized states), thus giving a very wide range of possible emission lines for each of over 70 elements. The large number of emission lines creates a problem of *spectral interference*, but it also gives the opportunity to measure the emission of a particular element at more than one wavelength.

The need to resolve a large number of close emission lines requires a mono-chromator system with a very high resolution. This has been achieved by using an echelle grating and a prism (see Section 16.3), together with a CTD detector (see Section 18.4).

DQ 6.5

Which factors could contribute to giving an ICP-AES system a wide dynamic range:

(a) the ability to change the integration period of a CTD detector;
(b) high resolution of the echelle grating;
(c) use of different emission lines from the same element?

Answer

The use of the CTD detector gives simultaneous recording across the range of the emission spectrum, and allows the simultaneous measurement of different elements, but it also allows the measurement of more than one emission wavelength from the same element. By choosing lines of different relative intensities, it is possible to record both very concentrated and diluted samples within the measurement range of the instrument. In addition, some forms (see Section 18.4) of the CTD detector allow the possibility of choosing the integration time of each wavelength region (pixel) to match the intensity of emission at that wavelength. This also extends the dynamic range of this type of detector.

The high resolution and accuracy of the echelle grating gives a very detailed spectral response, which helps to separate the wanted analyte signal from other close, and overlapping, emission wavelengths. In addition, by using the very high resolution echelle grating system, it is possible to monitor the background radiation at either side of the emission line and make an interpolation for the background radiation at the wavelength of the analyte.

The wide dynamic range of the ICP-AES system, coupled with the simultaneous measurement of multiple emission lines from most elements, means that it is possible to analyse rapidly for many elements within the same sample.

6.5 Performance Characteristics

The response of the system is often defined by the **characteristic concentration** (or *characteristic mass*) which gives a 1% absorption of radiation.

DQ 6.6

A sample absorbs 1% of the incident radiation. Calculate its absorbance.

Answer

If it absorbs 1%, then its transmittance is 99%, i.e. $T = 0.99$. *The absorbance,* A, *can then be calculated from* $A = -\log T$, *as follows:*

$$A = -\log(0.99) = 0.004\,36.$$

Atomic spectroscopy is often used to measure elements at trace levels, and consequently the *detection limit* is an important performance consideration. The *instrumental* limits to detection are mainly due to instrument noise and the imprecision of atomization.

The concept of signal-to-noise ratio is well defined for F-AAS and for ICP-AES in that a *constant* analytical signal must be distinguished from random noise. The noise signal can be reduced by using either of the following approaches (see Section 14.5):

- signal integration;
- a low-pass filter.

The signal-to-noise ratio can be increased by averaging the signal over an increased integration time, T_{int}. In this mode, the integrated value is only displayed after the integration time, and remains constant over the next integration period until the new averaged signal is displayed.

A low-pass filter will also improve the signal-to-noise ratio by smoothing the noise. In this mode, however, the changing signal will be displayed continuously, but with the random noise fluctuations dampened.

DQ 6.7

Which noise-reducing mode, i.e. integration or filter, should be used when carrying out the following:

(i) tuning the instrument to optimize the absorption signal;
(ii) taking the reading?

Answer

When tuning the instrument (adjusting the wavelength or flame height), it is important to see the way in which any adjustment is affecting the result. In the integration mode, any change in output is not seen until after the integration period, thus making it very difficult to make interactive adjustments. The low-pass filter is more suitable for this task.

However, when taking the final reading, it is important to get a single 'averaged' result, and it is convenient to use the integration mode to do this.

The detection limit of the method is also dependent on the ability to provide valid reproducible 'blanks'. This is a particular problem for GF-AAS where the volatized products of the sample matrix produce significant background absorbance. Thus, detection limits are very dependent on matrix reproducibility, and, with such practical variability, it is not appropriate to quote detection limits.

The other main specifications of the AAS systems relate to the performance characteristics of the monochromator (see Chapter 16), as follows:

Spectral bandwidth. A narrow bandwidth reduces the range of flame emissions that can become included with the measurement. For GF-AAS, it is also beneficial to reduce the height of the slit due to the small volume over which the absorption occurs.

Focal length and f-*number.* The *f*-number of a focusing system is defined as the focal length of the system, *f*, divided by the diameter, *d*, of the lens (or mirror). For example, typical *f*-numbers for a photographic camera are 2.8, 5.6, 11 and 22. A low '*f*-no.' gives greater energy throughput from the monochromator, but it also gives a more divergent beam, which is a disadvantage when focusing the beam along the plane of the flame in the F-AAS system.

SAQ 6.3

Explain the meaning of the following specification for an atomic absorption spectrophotometer:

Linearity > 0.75 A *with precision of* < 0.5% (RSD) *from 10 integrations of 5 s for 5 mg/l Cu.*

SAQ 6.4

ICP-AES and ICP-MS both suffer from 'spectral' interferences. Compare these two types of interferences.

Note. Further questions that relate to the topics covered in this chapter can be found in SAQs 3.5, 12.3 and 16.4, and DQs 1.3, 3.7, 3.8, 13.12, 14.10, 16.2, 16.5, 17.3 and 18.4.

Summary

This chapter has looked at those factors that influence the result of atomic spectroscopy measurements. It differentiated between the three main types of atomization process — flame, furnace and plasma, and identified the performance characteristics used to define the performance of such instruments.

Certain factors were highlighted, including the limited dynamic range of the absorption systems, the problem of defining detectability for the furnace system, and the abundance of spectral lines in the plasma system. The problems of background absorption and the three main methods of its correction were also introduced.

Reference

1. Smith, S. B. and Hieftje, G. M., *Appl. Spectrosc.*, **37**, 419–424 (1983).

Chapter 7

Fourier-Transform Infrared Spectrophotometers

Learning Objectives

- To describe the design of an FTIR spectrophotometer.
- To appreciate why information about fine detail in the wavenumber spectrum is distributed across a wide range in the interferogram, and how broad details are encoded near to the centre of the interferogram.
- To quantify the relationship between resolution and length of mirror travel.
- To understand the process of apodization and its effect on spectral linewidth.
- To identify the performance characteristics appropriate to the FTIR spectrophotometer.

7.1 Basic System

An infrared spectrophotometer records the transmittance, T (or absorbance (A)) of a Test Sample for wavelengths within the mid-infrared (IR) region of the electromagnetic spectrum (400 to 4000 cm^{-1}). The IR spectrum is normally drawn as a function of the 'wavenumber', σ, which is defined as 'the reciprocal of wavelength, measured in centimetres' (see Section 13.6), as follows:

$$\sigma = 1/\lambda \qquad (7.1)$$

Until about 1990, IR spectrophotometers used *wavelength-dispersive* systems with a scanning monochromator, with the latter being very similar in design to that used in UV–visible spectrophotometers.

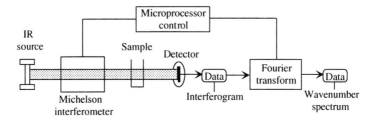

Figure 7.1 Schematic representation of a Fourier-transform infrared spectrophotometer.

However, the performance of these scanning IR systems was severely limited by the performance capabilities of the thermal IR detectors available at the time, which were slow in comparison to photon detectors (see Section 18.1) and had poor signal-to-noise (S/N) characteristics. The development of faster IR detectors, plus the use of Fourier-transform (FT) systems has permitted the use of a totally different method for collecting spectral data.

Figure 7.1 shows a schematic layout of a Fourier-transform infrared (FTIR) spectrophotometer. It is important to realize that the FTIR system does not 'scan' the spectrum by moving sequentially from *wavelength to wavelength* in the same way as a conventional spectrophotometer. A typical FTIR instrument uses a *Michelson Interferometer* instead of a monochromator.

Figure 7.2 gives a simple comparison between the transmittance profile of a monochromator with that of an idealized Michelson interferometer. A monochromator only transmits a very narrow band of wavenumbers at any given setting of the diffraction grating. In contrast, the interferometer has a *sinusoidal* transmittance pattern, T_δ, which is spread across the full spectral range.

The signal recorded by the detector of the spectrophotometer is the *sum* of the radiation, at *all wavenumbers*, that has passed through both the interferometer and the sample.

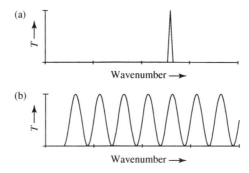

Figure 7.2 Comparison between the transmittance profiles obtained from (a) a monochromator and (b) an idealized Michelson interferometer.

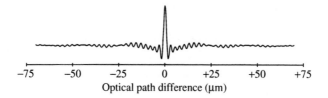

Figure 7.3 An idealized interferogram — a plot of detector signal against optical path density.

The interferometer contains a movable mirror, which changes the optical path difference (opd), δ, between two beams of light. A change in the opd changes the cycle length of the sinusoidal pattern — see below in Figure 7.5(a). A fuller discussion of the operation of the Michelson interferometer is given in Section 7.2.

The detector signal is recorded for different values of the optical path difference (δ) in the interferometer. A plot of detector signal against optical path difference is called an Interferogram (Figure 7.3).

At each value of the opd in the interferogram, the magnitude of the aggregate signal depends on how well the sinusoidal pattern in the radiation 'matches' the transmittance of the sample across its spectrum. The transmittance profile of the sample is therefore *encoded* in the interferogram in a complex way.

The interferogram can not be interpreted directly by the analyst and must first be analysed mathematically. It can be shown (see Section 7.4) that the interferogram is the inverse Fourier transform (see Section 13.4.1) of the wavenumber spectrum. It is then possible to use a microprocessor system to carry out a Fourier transform to regenerate the conventional spectrum for use by the analyst. This process has therefore given the name to this new breed of instruments, namely the Fourier-Transform Infrared (FTIR) Spectrophotometer.

7.1.1 Background Spectrum

The typical FTIR spectrophotometer is a single-beam instrument. Before recording the intensity, $I(\sigma)$, of the *test sample*, it is necessary to record the reference background intensity, $I_0(\sigma)$, for all wavenumbers. This takes account of the variation of source intensity across the spectrum, together with the absorption of radiation by carbon dioxide and water vapour present in the air. The value of the transmittance for each wavenumber, $T(\sigma)$, is given by the following expression:

$$T(\sigma) = \frac{I(\sigma)}{I_0(\sigma)} \tag{7.2}$$

7.2 Michelson Interferometer

If there is no diffraction grating to rotate, which component in the interferometer moves?

The operation of an FTIR system can be most easily understood by examining the Michelson interferometer, which is at the heart of the system (Figure 7.4).

In the diagram of the Michelson interferometer, you should notice the following:

- there is *no diffraction grating* in the system;
- there is *no selection* of particular wavelengths (or wavenumbers);
- the sample is, at all times, exposed to the *full spectral range* of radiation.

The radiation from the IR source is directed as a parallel beam on to the beam-splitter, which is a coated plate that partly transmits and partly reflects the radiation. This divides the incoming radiation into two beams, with approximately half of the energy being reflected towards a fixed mirror (F), and the other half being transmitted towards a movable mirror (M).

The transparency of the material from which the beam-splitter is made is a crucial factor in determining the *wavenumber range* over which the instrument can operate. The commonly used material, KBr, allows transmission of IR radiation in the range 370 to $10\,000$ cm^{-1}, while other materials permit other operational ranges.

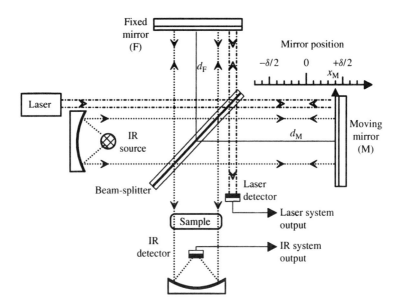

Figure 7.4 Schematic representation of a Michelson interferometer.

The two beams are reflected by the flat mirrors back towards the beam-splitter which again partly reflects and partly transmits the radiation. The result is that approximately half of each beam eventually reaches the detector, although having travelled different paths.

In Figure 7.4, the position of the movable mirror (M) is shown by x_M, while the distances from the centre of the beam-splitter to the fixed mirror and the movable mirror are given by d_F and d_M, respectively. We can then define the position of the mirror as being at 'zero' ($x_M = 0$) when the system is symmetrical, as follows:

$$x_M = d_M - d_F \qquad (7.3)$$

If the position of the mirror M changes, one beam will travel a different distance, and there will be an *optical path difference*, δ, (also called the *retardation*) between the two beams, given by the following expression:

$$\delta = 2(d_M - d_F) = 2x_M \qquad (7.4)$$

7.2.1 Transmittance of the Michelson Interferometer

If the transmittance versus wavenumber plot for the interferometer is sinusoidal, which factor determines the cycle length?

When the two beams of the Michelson interferometer reach the detector, 'wave interference' occurs. The electromagnetic fields of each of the two beams will add or subtract depending on their relative phases.

The phase difference, *measured in number of cycles*, between the two beams is given by the following:

$$\text{Phase Difference (in cycles)} = \frac{\textit{Different} \text{ distance travelled } (= \text{opd}, \delta)}{\text{Length of each cycle } (= \text{wavelength}, \lambda)} \qquad (7.5)$$

One complete cycle is equal to a phase difference of 2π. Hence, the *phase difference, φ (in radians)* between the two beams is given by the following:

$$\varphi = 2\pi\delta/\lambda = 2\pi\delta\sigma \qquad (7.6)$$

If either (i) the two beams travel equal distances ($\delta = 0$), or (ii) the path difference is equal to an *integer* number of wavelengths ($\delta = \lambda$, 2λ, 3λ, etc.), then the phase difference will be $n \times 2\pi$ (where n is an integer). In this case, the beams will arrive at the detector 'in phase' (or in step) and will add up to give *maximum intensity*.

If, however, the optical path difference is equal to an integer number of wavelengths, *plus one half* wavelength ($\delta = 0.5\lambda$, 2.5λ, 3.5λ, etc.), the beams will arrive at the detector 'out-of-phase'. They will then interfere destructively, thus giving zero intensity and zero transmission through the Michelson interferometer.

The transmittance of the interferometer will therefore depend on the phase difference, φ, and hence on the product, $\delta\sigma$, according to the following equation:

$$T_{\delta\sigma} = \tfrac{1}{2}(1 + \cos\varphi) = \tfrac{1}{2}[1 + \cos(2\pi\delta\sigma)] \tag{7.7}$$

The wavenumber plots, (a(i)), (a(ii)), (a(iii)) and (a(iv)), presented in Figure 7.5 show the relative transmittance profiles $(T_{\delta\sigma})$ of the Michelson interferometer calculated for opd values equal to 3.00, 4.50, 33.00 and 34.25 μm, respectively. Each value of the opd corresponds to a particular position of the movable mirror. For each of these positions, the plots show the transmittance as a *function of wavenumber*, σ. Every point on each plot in Figure 7.5 is calculated from equation (7.7) by using appropriate values of δ and σ.

At each different position of the movable mirror (different opd), the interferometer produces a different spectral profile of radiation. The *transmittance profiles* of the interferometer are all sinusoidal in wavenumber, but the *cycle length decreases* with *increasing opd*.

DQ 7.1

Calculate the cycle length of the transmittance profile of the interferometer at an optical path difference of -200 μm.

Answer

For a complete cycle, the value of $2\pi\delta\sigma$ must change by $\pm 2\pi$. Thus, the cycle length, $d\sigma$, must have a value such that $\delta \times d\sigma = \pm 1$. Hence, $d\sigma = \pm 1/\delta = \pm 1/(200 \times 10^{-6}) = \pm 5000 \ m^{-1} = \pm 50 \ cm^{-1}$.

Each cycle will correspond to a change in wavenumber, $d\sigma (= 50 \ cm^{-1})$.

NB — The negative value for the opd only indicates that the movable mirror is closer to the beam-splitter than the fixed mirror, but, as the interferogram is symmetrical about the 'zero' position, this does not change the numerical result of this question.

The power (P_δ) recorded at the detector for a particular value of the opd, δ (i.e. the mirror position), will be given by the following:

$$P_\delta = \int_{\sigma_{min}}^{\sigma_{max}} P_\sigma \times T_{\delta\sigma} \times T_{S\sigma} d\sigma \tag{7.8}$$

This is the summation, across all wavenumbers in the spectral range (from σ_{min} to σ_{max}), of the product of P_σ, the spectral power output of the source, $T_{\delta\sigma}$, the transmittance of the interferometer, and $T_{S\sigma}$, the transmittance of the sample, all at wavenumber, σ.

Note that the energy falling on the detector includes contributions from *all wavenumbers* in the spectrum.

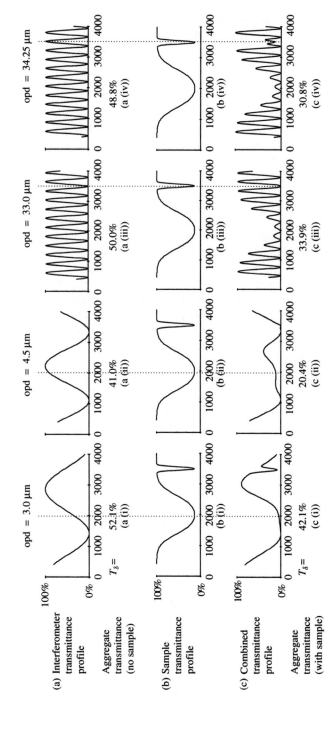

Figure 7.5 Transmittance profiles obtained from the Michelson interferometer: (a) interferometer (no sample) profile; (b) sample profile; (c) combined profile (abscissae are given in wavenumbers (cm^{-1})).

The plot of P_δ for different values of the opd, δ, is the *interferogram*. An idealized example of the latter is shown above in Figure 7.3.

The transmittance, $T_{\delta\sigma}$, of the Michelson interferometer depends on both the retardation, δ, and the wavenumber, σ. This gives us two possible approaches to understanding the information contained in the interferogram, as follows:

> *Approach I.* For *each value of the opd*, δ, we look at the spectral transmittance across the wavenumbers, σ, and calculate the aggregate transmittance by integrating across the spectrum. This gives us *one point* in the interferogram. We then change the opd to generate the rest of the points for the complete interferogram (Section 7.3).

> *Approach II.* We derive each 'interferogram component' (cf. frequency components, see Section 13.4) that would be produced by every *wavenumber component*. We then add all of these separate interferograms together to create the final interferogram (Section 7.4).

Each approach has its particular merits and tells us the same story in a different way. Approach I may be useful for developing an *intuitive* understanding of the system, while Approach II derives some useful mathematical relationships.

7.2.2 Use of Laser Light

The purpose of the He–Ne laser used in FTIR systems is to provide an optical scale against which to measure distances in the interferogram. The laser light passes through a parallel interferometer system (see Figure 7.4) which uses the same mirrors as the IR system, and produces an interferogram through a separate detector.

Due to the fact that the spectral bandwidth of laser light is very narrow and has very good coherence (see Section 17.3.4), its interferogram is a simple cosine wave extending over the full length of the mirror travel. This cosine wave is then used as a spatial grid with an exactly known scale, which thus enables the position of the mirror to be monitored accurately.

7.3 Interferogram — Approach I

7.3.1 Without Sample

In order to understand the interferogram, we will first consider the operation of the system without a sample, i.e. put $T_{S\sigma} = 1$ in equation (7.8) for all values of σ. We will also make a simplifying assumption that the source of radiation provides a spectral output, P_σ, which has the same value, P, between 400 and 4000 cm^{-1}, and is zero at all other wavenumbers.

Equation (7.8) can then be written as follows:

$$P_\delta = P \int_{400}^{4000} T_{\delta\sigma}\, d\sigma \qquad (7.9)$$

The aggregate *transmittance* (at opd = δ) is then given by the following:

$$T_\delta = \frac{P_\delta}{P} = \int_{400}^{4000} T_{\delta\sigma}\, d\sigma \qquad (7.10)$$

This is the summation over all wavenumbers for the particular value of the opd (δ).

The values of the aggregate transmittance, T_δ, are given in Figure 7.5 for each of the specific values of the opd, i.e. $T_\delta = 52.1, 41.0, 50.0$ and 48.8%, for opd values of 3.0, 4.5, 33.0 and 34.25 μm, respectively. A complete plot is shown later in Figure 7.6 (plot(a)) for T_δ, without a sample, for values of the opd up to 50 μm.

There is always a burst of energy at zero opd ($\delta = 0$), corresponding to the symmetrical position of the interferometer, where the interference is constructive for all wavenumbers.

Even without an absorbing sample, the interferogram shows a complex oscillatory shape. This is actually the superposition of two sinc functions (see Section 7.4.3), caused by the fact that the spectral energy in our example is cut off abruptly at both ends (400 and 4000 cm^{-1}) of the spectrum. The short oscillation with a cycle length of 2.5 μm is due to the fact that the spectrum is truncated at 4000 cm^{-1} (note that 2.5 μm = 1/4000 cm^{-1}). There is also a longer oscillation which can only just be observed as a low amplitude wave with a cycle length of 25 μm due to the cut-off at 400 cm^{-1} (note that 25 μm = 1/400 cm^{-1}).

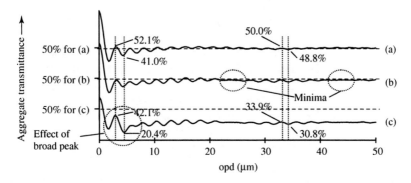

Figure 7.6 Development of an idealized interferogram, showing the effect of a sample: (a) no absorbing sample; (b) absorbing sample, with only the narrow line; (c) absorbing sample, with both the broad and narrow lines.

7.3.2 With Sample

We can 'test' for the effect of a sample by using an *idealized test sample*, with the transmittance profile shown (repeatedly) as (b(i)) to (b(iv)) in Figure 7.5. This profile shows the following:

- a broad absorption line at 2000 cm^{-1};
- a narrow absorption line at 3500 cm^{-1}.

The energy profile reaching the detector will depend on the product of the transmittance values in plots (a) and plots (b) in Figure 7.5. The combined transmittance profiles are shown in this figure as plots (c(i)) to (c(iv)). The new values of the aggregate transmittance are given as $T_\delta = 42.1$, 20.4, 33.9 and 30.8%, for opd = 3.0, 4.5, 33.0 and 34.25 μm, respectively.

The transmittances shown in plots (c) depend on the relative positions of the maxima and minima in plots (a) and (b). For example, in *both* (a(i)) and (b(i)) good transmittance occurs at about 3000 cm^{-1} and below 1000 cm^{-1}, and the result is a moderately high value (42.1%) of the *aggregate* transmittance. However, the minima in (a(ii)) and (b(ii)) occur at different wavenumbers with the effect of reducing the combined transmittance over a wide range of wavenumbers and so giving a low value (20.4%) for the aggregate transmittance.

DQ 7.2

Explain why the plots (c(iii)) and (c(iv)) in Figure 7.5, for opds of 33.00 and 34.25 μm, respectively, show significant differences in shape at about 3500 cm^{-1}, and why this gives an *additional* drop in energy for (c(iv)).

Answer

*At 3500 cm^{-1}, the sample has a **minimum** transmittance. If this coincides with a minimum in the interferometer profile (as in (iii)), then the overall transmittance is hardly affected.*

If, however, the minimum transmittance of the sample coincides with a maximum in the interferometer profile (as in (iv)), the overall transmittance at 3500 cm^{-1} is dramatically reduced, and the aggregate signal value is also reduced. Note that the difference between (a(iii)) and (a(iv)) is only 1.2% (50.0 − 48.8), whereas the difference between (c(iii)) and (c(iv)) is 3.1% (33.9 − 30.8), i.e. the effect of the narrow line has accentuated the oscillation at around 34 μm (see Figure 7.6(c)).

7.3.3 Information in the Interferogram

Each point in the interferogram contains *different* information about the transmission spectrum of the sample.

Figure 7.6 shows the development of an idealized interferogram produced by using the following:

(a) no absorbing sample;
(b) absorbing sample, as in Figure 7.5(a), but with only the *narrow* line;
(c) absorbing sample, as in Figure 7.5(a), with *both* the broad and narrow lines.

By careful comparison of the plots shown in Figure 7.6, we can observe the following main features:

(i) The effect of the narrow absorbing line appears as a *modulation* of the oscillations in the interferogram with the *minima* shown in the figure. By comparing plots (a) and (b), it can be seen that the oscillations increase in amplitude in the region of 34 μm — see DQ 7.2 for the reason for this. Note that the effect of the *narrow* line produces a *broad* response in the interferogram.

(ii) The effect of the *broad* peak, however, is more noticeable at *low* values of the opd when the oscillatory cycle of the interferometer transmittance is more closely matched with that of the sample transmittance.

In general, the information about *broad* structures in the sample spectrum is contained close to the *centre* of the interferogram, while the information about *fine* structure extends to *larger* values of the optical path difference.

It should be noted that the interferograms shown in Figure 7.6 are idealized in that the IR source is assumed to be uniform between 400 and 4000 cm^{-1}. In a real system, the effect of the source intensity profile is taken into account by running the background spectrum *without the sample* before making the test measurement.

An additional difference that can also be observed between the real interferogram and the idealized version is that the real plot is often *not symmetrical* about zero path difference. This is due to wavelength-dependent phase shifts in the beam-splitter, and is an additional correction factor that the computerized signal-processing facility must accommodate.

7.4 Interferogram — Approach II

Is there an easy way of interpreting the information in the interferogram?

7.4.1 Interferogram and its Fourier Transform

In Approach II, we will carry out the following:

(i) develop the interferogram *components* corresponding to specific values of the wavenumber σ, i.e. $T_{\delta\sigma}$ as a function of δ;

(ii) add these components together to obtain the final interferogram.

Consider a theoretical sample which only transmits radiation at 'one' wavenumber, i.e. σ_A. Equation (7.7) shows that the signal, plotted against the opd in the interferogram, will be a cosine wave with a cycle length of $1/\sigma_A$, and is given by the following expression:

$$T_{\delta\sigma} = \tfrac{1}{2}[1 + \cos{(2\pi\delta\sigma_A)}] \tag{7.11}$$

This is illustrated in Figure 7.7(a) for a monochromatic signal at 500 cm^{-1}, and in Figure 7.7(b) for a monochromatic signal (of lesser amplitude) at 540 cm^{-1}

Each spectral component (of wavenumber σ_A) will result in a cosine-wave component in the interferogram. The cycle length of this component (in opd) will be equal to the wavelength, $\lambda_A (= 1/\sigma_A)$, of the component. The full spectrum of many frequency components will give a combined interferogram made up of many cosine components.

Each component in the wavenumber (cm^{-1}) domain is equivalent to a cosine wave in the opd (distance) domain. Section 3.4.1 describes how a sine (or cosine) wave in one domain (e.g. time) can be represented by a single component in a conjugate domain (e.g. frequency), and that a complex signal in the first domain can be described by a distribution of components in the second domain. The same situation occurs with the interferogram, which can also be expressed as the sum of cosine components, each of which corresponds to a single wavenumber component in the IR spectrum.

*Thus, the transmittance spectrum, T_σ, is the **Fourier transform** of the interferogram, T_δ.*

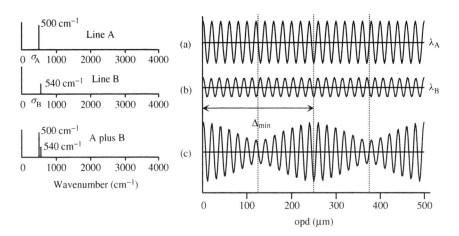

Figure 7.7 Resolution in the interferogram, shown for: (a) line A (500 cm^{-1}); (b) line B (540 cm^{-1}); (c) line A plus line B.

While the mathematical basis of the Fourier transform has been well known for a long time, the *general* process for transforming a typical spectrum is a long and involved calculation, even for a microprocessor. The need to speed up the calculation was met with the development of a Fast Fourier-Transform (FFT) algorithm which allows a rapid calculation based on the repeated pairing of data values. The speed of the FFT can only be achieved if the number of data points in the spectrum is equal to 2^N, where 'N' is an integer. This FFT can only be performed if the interferogram is digitized by using 2^N data points (e.g. 4096, 8192, etc.).

7.4.2 Resolution

We can use Approach II to discover which factor limits the resolution of the FTIR spectrophotometer. In a spectrum, the 'resolution' is quantified as the minimum separation between two lines (wavelength or wavenumber) at which they can still be seen as two separate lines.

For the FTIR system, we need to consider how the lines are 'resolved' in the *interferogram*.

Consider two lines (A and B) close together in an IR spectrum, as shown in Figure 7.7. These have the wavenumbers, $\sigma_A (= 500 \text{ cm}^{-1})$ and $\sigma_B (= 540 \text{ cm}^{-1})$, respectively, with $\sigma_A < \sigma_B$. The equivalent wavelengths (in cm) are $\lambda_A = 1/\sigma_A$ and $\lambda_B = 1/\sigma_B$, with $\lambda_A > \lambda_B$. We know from equation (7.7) that these are also the cycle lengths of the component waves in the interferogram.

Since the magnitudes of λ_A and λ_B are very similar, the two waves may take many cycles in the interferogram before they become *out-of-phase* and produce destructive interference in the interferogram.

The resulting wave in the interferogram has an 'envelope' which oscillates between the maximum and minimum amplitudes. The cycle length of the envelope is shown as Δ_{min}, and this corresponds to the distance in opd for which the phase difference between the two waves A and B increases by 2π.

The criterion for resolving A and B is that the instrument must be able to record *one full cycle* of the envelope in the interferogram. In practice therefore, the *mirror must be able to travel a minimum distance* which is equivalent to Δ_{min} in *optical path difference*. Since the optical path difference created by a mirror movement is twice the *physical* movement of the mirror, the actual distance travelled by the mirror will be $\Delta_{min}/2$.

The parameter Δ_{min} is the distance in opd over which a phase difference of 2π develops between A and B. If there are n cycles for wave A within the distance Δ_{min}, then there must be one more cycle, i.e. $n+1$, for wave B (remember $\lambda_A > \lambda_B$), as follows:

$$n \times \lambda_A = \Delta_{min} \quad \text{and} \quad (n+1) \times \lambda_B = \Delta_{min} \quad (7.12)$$

By eliminating 'n' from these two simultaneous equations, we obtain:

$$\Delta_{min} = \lambda_A \lambda_B / (\lambda_A - \lambda_B) = 1/(1/\lambda_B - 1/\lambda_A) = 1/(\sigma_B - \sigma_A) \quad (7.13)$$

Thus, the minimum resolution is given by the following expression:

$$(\sigma_B - \sigma_A) = 1/\Delta_{min} \tag{7.14}$$

The resolution of the FTIR spectrophotometer $(\sigma_B - \sigma_A)$ is determined by the physical length of travel $(\Delta_{min}/2)$ of the mirror in the Michelson interferometer. An increase in the mirror travel distance will thus allow resolution of those lines that are closer together.

DQ 7.3

Calculate the minimum distance that a mirror should travel if the instrument should be able to resolve two lines at 792 and 796 cm^{-1}

Answer

By using equation (7.14), the length of the interferogram required to record the necessary information will be $\Delta_{min} = 1/(796 - 792) = 1/4 = 0.25$ cm. This requires the mirror to travel at least the minimum physical distance, $\Delta_{min}/2 = 0.125$ cm.

It is a peculiar consequence of the Fourier transform that *resolution* in a conventional spectrophotometer is determined by how *small* the slitwidth can be made in the monochromator, while for the FTIR system the *resolution* depends on how *large* it is possible to make the range of mirror travel.

SAQ 7.1

Estimate the minimum time that it would take an FTIR spectrophotometer to record a single spectrum with a resolution of 0.5 cm^{-1} if the mirror travels with an opd velocity of 0.5 cm/s.

7.4.3 Instrument Lineshape

The instrument lineshape (ILS) of a spectrophotometer is *the limiting lineshape, due to the instrument*, that would be produced in the output spectrum in response to a theoretical absorption line which has a *zero* spectral width.

The origin of the ILS can be understood by considering the response of the Michelson interferometer to an input spectrum with a single *monochromatic* (theoretical) line, as shown in Figure 7.8. This figure shows the 'wavenumber' domain (on the right) and the 'opd' domain (on the left). The Fourier-transform process (see Section 13.4.1) converts a signal in the opd domain into the equivalent signal in the wavenumber domain.

The interferogram for a *monochromatic* input will be a pure sine wave (Figure 7.8(a)). If the sine wave *extended to infinity* (i.e. infinite mirror travel), then its Fourier transform would also be a single wavelength component.

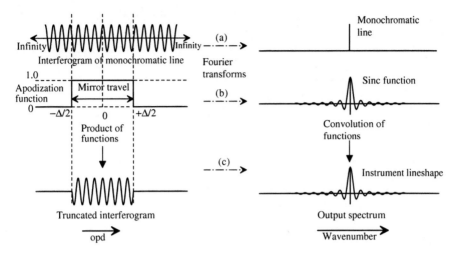

Figure 7.8 Illustration of instrument lineshape, showing the response of the Michelson interferometer to an input signal to give the final output signal.

However, the mirror has only a finite distance, Δ, to travel. The actual (truncated) interferogram is the *product* of the infinite sine wave, multiplied with a rectangular 'top-hat' function (Figure 7.8(b)). This 'top-hat' function has a value of '1' within the length of mirror travel and a value of '0' everywhere else. We now need to work out the Fourier transform of the new *truncated* interferogram.

An important mathematical fact to note (see Section 15.3.1) is that:

'the Fourier transform of a *product* of two functions'

is the same as

'the *convolution* of the Fourier transforms of the *individual* functions'.

It is also a fact that the Fourier transform of the top-hat function is a sinc x function[†] (see Figure 7.8(b)).

The convolution of any function with a *single 'line' function* produces the *original function*. Hence, the convolution of the single 'line' with the sinc function provides an output line with a sinc-function lineshape. Thus, the Fourier transform of our *truncated* sine wave is a *sinc function* as shown.

This sinc x function has the following features:

(i) a central peak;
(ii) significant side wings;
(iii) some negative values.

[†] Sinc function: sinc $x = (\sin x)/x$.

This peculiar shape does not have real *physical* significance, and is due entirely to the limitations of mirror travel and the mathematics of the transform process. However, this shape is inconvenient as it will both distort and broaden the lines in the observed spectrum.

The *width* of the sinc function is inversely proportional to the width of the top-hat function, Δ_M (equal to the length of the interferogram). Thus, the width of the instrument lineshape in the spectrum is inversely proportional to the length of mirror travel, as follows:

$$\Delta\sigma \propto \frac{1}{\Delta_M} \qquad (7.15)$$

This is consistent with equation (7.14), in that the minimum separation ($\sigma_B - \sigma_A$) between two resolvable lines is also inversely proportional to Δ_M.

7.4.4 Apodization

The sinc σ shape of the ILS, as described above, arises 'mathematically' from the top-hat function produced by simply truncating the sine-wave interferogram. The sinc σ shape is particularly inconvenient because it has large side lobes and some apparent *negative* values for transmittance.

It is possible to produce different lineshapes by performing *additional* 'mathematical' processing of the interferogram data. This process is known as Apodization.

For simplicity, we will describe below a process where it is possible to produce the sinc^2 σ instrument lineshape shown in Figure 7.9. Note, however, that although the sinc^2 σ function is actually broader than the sinc σ line for

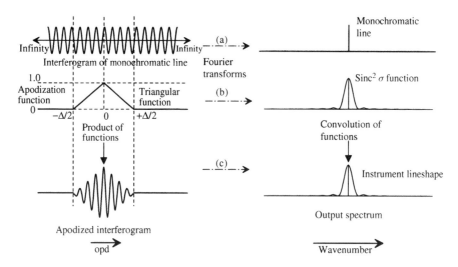

Figure 7.9 Illustration of triangular apodization.

the same mirror travel, the side lobes of the $\text{sinc}^2 \, \sigma$ function are proportionately much smaller and there are no negative values.

Each data point, δ_i, in the interferogram is multiplied by a value, a_i, derived from an apodization function.

Before we introduce a 'new' apodization, it is useful to remember that the effect of truncating the mirror travel is itself equivalent to multiplying the interferogram data by an apodization function. This 'top-hat' function is given by the following:

- $a_i = 1$ for points, δ_i, within the range of mirror travel, Δ;
- $a_i = 0$ for all other points.

The apodization function required to give a $\text{sinc}^2 \, \sigma$ shape is a 'triangular' function described by the following:

- $a_i = 1$, for $\delta = 0$;
- $a_i = |\delta|/\Delta$, for $-\Delta/2 < \delta$ and $\delta < +\Delta/2$;
- $a_i = 0$, for $-\Delta/2 > \delta$ and $\delta > +\Delta/2$.

The Fourier transform of a triangular function is a $\text{sinc}^2 \, \sigma$ function, and this will become the instrument lineshape. Notice that the negative side lobes have disappeared — the name, apodization, actually derives from its the Latin meaning, i.e. 'cutting off the feet'!

There are different apodization functions that can be used. However, any apodization to improve the *lineshape* will also have the effect of *broadening* the line. This is because lineshape is improved by reducing the importance of the data points near the end of the mirror travel, but, as we have seen in Section 7.4.2, good resolution (i.e. narrow lines) requires data from the ends of the mirror travel in order to identify the envelope in the interferogram.

There is a trade-off between *linewidth* and *lineshape*, and the operator has a choice of stressing different characteristics in the data by choosing different apodization functions.

7.5 Advantages of FTIR Spectrophotometry

In a normal spectrophotometer, the different *wavelengths* within the spectrum are recorded *sequentially*. However, this 'scanning' process takes time. As there is no monochromator in an FTIR spectrophotometer, all wavelengths in the input spectrum fall on the detector simultaneously, and so are recorded simultaneously. The time taken to record the spectrum is therefore only a matter of seconds.

DQ 7.4

What is the value in obtaining the data within a shorter time?

Answer

The time taken to collect data is one of the most important parameters in the efficiency of any instrument system. Indeed, in some cases it is the time itself that is important, e.g. in analysing a sample in an 'on-line' production process, or in the simultaneous detection and analysis of the output from a chromatograph.

Even if the time itself is not so important, it is still possible to use the extra 'time' gained when using an FTIR system by repeating the measurement several times and then averaging the result to obtain a better signal-to-noise ratio (see Section 14.5.5).

We have seen in the above discussion question that one of the advantages of fast measurement is that it may be possible to repeat the measurement '*m*' times and then average the results to obtain a better signal-to-noise ratio. The following relationship applies:

$$S/N \propto \sqrt{m} \tag{7.16}$$

However, there will only be a net advantage provided, of course, that S/N is not reduced by other factors. It is necessary to ask, for example, whether the fact that the increase in the radiation energy (i.e. the full spectrum) falling on the detector increases the *noise* produced by the detector.

DQ 7.5

Does an increased signal level increase the noise in a detector, if the type of noise is mainly:

(a) thermal noise;
(b) shot noise?

Answer

Referring to Section 14.4.3, it can be seen that thermal noise is independent of signal level, although this is not true for shot noise.

As thermal noise is the main source of noise in IR detectors, the increased signal level does not increase the noise (see DQ 7.5), and there is a net overall gain in S/N for FTIR systems. Note that this does not apply to UV–visible FT systems, where the 'time' gain is offset by increased noise.

DQ 7.6

When performing repeated scans, which of the following would be 'averaged':

(a) the interferograms;
(b) the final spectra?

Answer

*The conversion from the interferogram requires the use of the FFT algorithm (Section 7.4.1) and takes some time. It is therefore a more efficient use of time to average the repeated interferograms as they are produced, and then make a **single** software conversion to a final spectrum.*

SAQ 7.2

The signal-to-noise specifications for three different FTIR spectrophotometers are quoted as follows:

(a) 3000:1 (peak-to-peak) S/N for 4 cm^{-1} for 4 s;
(b) 12 000:1 (peak-to-peak) S/N for 4 cm^{-1} for 1 min;
(c) 15 000:1 (rms) S/N for 4 cm^{-1} for 4 s.

Explain the meaning of the specifications and compare the different values.

7.6 Performance Characteristics

The tests required to monitor the performance quality of an FTIR spectrophotometer usually include the following:

- Radiation energy at detector;
- Noise;
- Wavenumber accuracy and precision;
- Photometric precision;
- Resolution.

The signal-to-noise ratio is an important factor in IR measurements, and the noise level in the output is monitored as part of a regular performance check. Furthermore, in addition to ensuring that the noise is reduced to a minimum, it is vitally important that the magnitude of the signal should be at a maximum.

The performance of the Michelson interferometer is critically dependent on having the mirrors exactly perpendicular to the optical axes of the two beams. If this is not the case, the energy of constructive interference at the detector is very significantly reduced. The various manufacturers offer different solutions to the problem, e.g. mechanical adjustment of the position of one of the two plane mirrors (either manually or automatically), or the use of prismatic mirrors (cf. Cat's-eyes in the road!) which automatically reflect the light correctly. Whatever system is used, it is important to monitor the available energy on a regular basis.

The standard wavelength and photometric characteristics are measured by using standard filters, e.g. polystyrene.

SAQ 7.3

Explain why a decrease in spectral bandwidth will result in a decrease in the signal-to-noise ratio (all other factors staying equal) for the following:

(i) a UV–visible spectrophotometer;
(ii) an FTIR spectrophotometer.

SAQ 7.4

Why is the signal-to-noise ratio an important performance characteristic for an FTIR spectrophotometer, and which performance tests are normally carried out in order to ensure that the noise is minimized?

SAQ 7.5

What is the function of the *apodization* control in an FTIR spectrophotometer?

Note. Further questions that relate to the topics covered in this chapter can be found in SAQ 4.5, and DQs 1.3, 13.6, 17.1, 17.4 and 18.5.

Summary

This chapter has described the main design features of a Fourier-Transform Infrared (FTIR) spectrophotometer, based on the use of the Michelson Interferometer.

It developed two separate approaches to help the understanding of the information content of the interferogram, and then used that understanding to explain why the interferogram has the complex shape that it does.

The advantages of the system were described in terms of its fast data acquisition, and the option of repeated scanning to provide enhanced signal-to-noise ratios through data averaging.

Chapter 8

System Characteristics of Gas Chromatography and High Performance Liquid Chromatography

Learning Objectives

- To differentiate between the various performance parameters used to describe chromatograms.
- To calculate the values of relevant parameters for a given chromatogram.
- To appreciate the choice of operating conditions available to optimize a particular separation.
- To identify the effects that the performance of specific instrument modules have on the system characteristics, and ultimately on the method characteristics.

8.1 Introduction

Is it possible to define the quality of a chromatogram **quantitatively?**

In this section, we will discuss the performance characteristics of a generic chromatographic separation. The specific systems and modular performances of High Performance Liquid Chromatography (HPLC) and Gas Chromatography (GC) are introduced in Chapters 9 and 10, respectively. The similar characteristics of the electropherograms in Capillary Electrophoresis (CE) are discussed in Chapter 11.

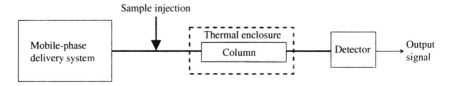

Figure 8.1 Schematic representation of the basic set-up for a chromatographic separation system.

The main elements of an instrumental separation system are illustrated in Figure 8.1. The analyte is injected into a *mobile* phase (liquid for HPLC or gas for GC), which is then forced under pressure though a 'column'.

For a given flow rate, the required *pressure difference* between the ends of the column will *increase* for the following:

- *longer* columns;

- *narrower* columns or more *tightly packed* columns;

- *increased viscosity* of the mobile phase (note that the viscosity of a liquid decreases with temperature, but increases for a gas).

The *flow* of the mobile phase can be measured, either in terms of the *linear velocity*, $\bar{u}$, or the *mass flowing per second*, as given by the following:

$$\text{Average linear velocity, } \bar{u} = \frac{\text{Length of the column, } L}{\text{Time to pass through column, } t_M} \quad (8.1)$$

$$\text{Mass flow rate} = \text{Linear velocity} \times \text{Fluid density}$$

$$\times \text{ Cross-sectional area} \quad (8.2)$$

DQ 8.1

For a stable gas flow **along a GC column**, identify how the value of each of the properties, (A), (B) and (C), listed in the table below, may change along the length of the column.

		Effect		
		(i)	(ii)	(iii)
Property	Parameter	No change	Increase	Decrease
A	Pressure			
B	Mass flow rate			
C	Linear velocity			

Answer

In a dynamic equilibrium, the mass entering each section of the column must equal the mass leaving that section. Hence, the mass flow rate is the same at all points throughout the system (including the detector) (B, (i)). The pressure drops along the column (A, (iii)), and the gas will expand. In order to maintain the same mass flow with an expanding gas (lower density), the linear velocity will increase (C, (ii)).

We see from the above discussion question that the linear velocity in GC increases along the column due to the expansion of the gas. For HPLC, the compressibility of the fluid is very small, and hence there is very little difference in linear velocity.

The viscosity of the gas in GC *increases* with increasing temperature. Hence, unless the pressure is *increased* during the run, the flow rate of the gas will tend to *decrease* with the rising temperature that occurs in a temperature programme.

The analyte sample is 'injected' as a short plug into the mobile phase, just before it enters the column. The column contains a *stationary* phase, with which the components of the analyte can interact. The stationary phase is held on an inert solid support material in the column — either on particle packing, or coated on the inside walls of capillary columns. If a component of the analyte has an affinity for the stationary phase, then the resulting interaction will delay the progress of that component through the column. The stronger the affinity, then the longer it takes for that component to emerge from the column.

A detector records the signal as each component arrives, thus producing the chromatogram. The *retention time* of each component is the time taken between the injection and its appearance at the detector. An illustrative chromatogram is shown in Figure 8.2.

The *aim* of a chromatographic method is to:

- separate the components of an analytical sample by achieving narrow peaks, with different retention times, and then to;
- enable a measurement of the area of each peak.

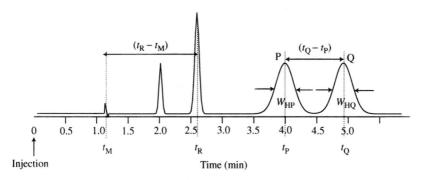

Figure 8.2 Illustration of a typical chromatogram.

The process of performance verification (e.g. through System Suitability Testing (Section 4.11)) requires that we can confirm that the *instrument* performance stays within acceptable limits. In the case of chromatography, we therefore need to be able to show that the chromatogram is sufficiently reproducible between runs. However, it is not sufficient to make a *subjective* judgement that two chromatograms *look* the same. We must identify the key characteristics of the chromatogram that are important in relation to the analytical process, and define relevant *quantitative* parameters that enable us to define the characteristics of the chromatograms with an appropriate set of numbers.

SAQ 8.1

Identify and compare the physical problems encountered in establishing constant flow rates in the following:

(i) a gas flow (GC);
(ii) a liquid flow (HPLC);
(iii) an electro-osmotic flow (CE).

8.2 Retention Time and Capacity Factor

Why is retention time less than ideal as a performance parameter?

The *retention time*, t_R, is the time interval between injection of the sample into the system and the arrival of the peak maximum at the detector.

In Figure 8.2, the first 'peak' is actually due to material that does *not* interact with the stationary phase in the column and therefore travels at the speed of the mobile phase. This is the unretained peak, whose retention time (void time) is given by t_M. The time spent by a component in the stationary phase is given by $(t_R - t_M)$ — the *adjusted* retention time.

The use of retention time as a universal performance characteristic is not ideal because of the following:

- it includes the time that the component has spent in the mobile phase (t_M), as well as the time spent in the stationary phase ($t_R - t_M$);
- it is a function of the flow rate of the mobile phase.

The *capacity factor (capacity ratio)*, k, gives the *fractional* delay introduced by the separation process. This is equal to the ratio of the times spent by the analyte in the stationary and mobile phases, respectively, as follows:

$$k = \frac{t_R - t_M}{t_M} \tag{8.3}$$

Small variations in the flow rate will have the same proportionate effect on both $(t_R - t_M)$ and t_M, and consequently this ratio is less sensitive to fluctuations in chromatographic conditions than the simple measure of retention time.

The capacity factor will be strongly affected by the following:

- oven temperature in GC;
- solvent composition in HPLC.

8.3 Peak Shape and Asymmetry

Can peak shape be quantified?

The actual shape for a symmetrical signal peak, S, can be represented by a Gaussian curve as a function of the abscissa, x, as given by the following:

$$S = \frac{1}{\sigma\sqrt{2\pi}} \exp\left[-\frac{(x - x_0)^2}{2\sigma^2}\right] \tag{8.4}$$

where x_0 is the position of the central maximum of the peak, and σ is the standard deviation of the peak around that maximum.

It is common practice to approximate the Gaussian curve with a simple triangle (see Figure 8.3), drawn by using tangents to the curve at the points of maximum slope (which occur at distances of σ from the maximum point of the curve).

The width of the triangle at half-height, W_H, and at the base of the triangle, W_B, are related to the standard deviation, σ, of the Gaussian peak as follows:

$$W_H = 2.355 \times \sigma; \quad W_B = 4 \times \sigma \tag{8.5}$$

The area of the triangle is 97% of the actual area of the Gaussian peak, but this difference is irrelevant, provided that *comparative* measurements of area are always made by using the *same* peak shape.

The *asymmetry* of an unsymmetrical peak (Figure 8.4) is defined in one of two ways, by using the ratio between the forward and trailing half-widths of the

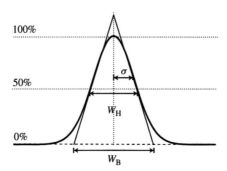

Figure 8.3 Approximation of the Gaussian shape of a chromatographic peak.

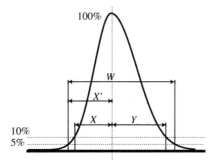

Figure 8.4 Illustration of the asymmetry of an unsymmetrical chromatographic peak.

peak. The tailing factor, T, is measured at 5% of the peak height, and the peak asymmetry, A, is measured at 10% of the peak height, as follows:

$$T_{5\%} = \frac{W}{2X'} \quad A_{10\%} = \frac{Y}{X} \qquad (8.6)$$

Dolan [1] has established an equivalence between the values of the tailing factor T_f and the values of the peak asymmetry, A. The results are presented in Figure 8.5.

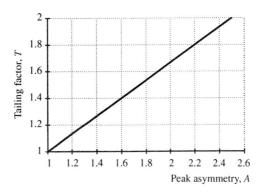

Figure 8.5 Tailing factor (T) as a function of the chromatographic peak asymmetry (A).

8.4 Column Efficiency

The term 'resolving power' can be used for both a column and a monochromator.

The *efficiency* of the column is the measure of its ability to produce a peak that is narrow (σ) in relation to its retention time (t_R). This is measured historically by the 'plate number', N, which is given by the following:

$$\text{(a) } N = \left(\frac{t_R}{\sigma}\right)^2 ; \quad \text{(b) } N = 5.545 \left(\frac{t_R}{W_H}\right)^2 ; \quad \text{(c) } N = 16 \left(\frac{t_R}{W_B}\right)^2 \quad (8.7)$$

Expressions (b) and (c) above are derived from expression (a) by squaring the ratios obtained from equation (8.5).

The 'resolving power' of the column, R_P, can be defined by using the ratio of *mean value* to *standard deviation*, as follows:

$$R_P = \left(\frac{t_R}{\sigma}\right) \quad (8.8)$$

Hence, we can then see that the resolving power is proportional to the square root of the plate number, as follows:

$$R_P = \sqrt{N} \quad (8.9)$$

Compare equation (8.8) with the resolving power of a mass spectrometer (see Section 12.1.1) and a monochromator (see Section 16.2.2).

DQ 8.2

Estimate the plate number for the chromatogram shown in Figure 8.2, by using the width of peak Q.

Answer

The width of peak Q is approximately 0.35 min and the retention time is approximately 4.95 min. This gives a plate number of $N \approx 5.55 \times (4.95/0.35)^2 \approx 1110$.

The plate number, N, is proportional to the length, L, of the column, so we can also define efficiency 'inversely' by using the following:

$$\text{Height of a theoretical plate (HETP)}, H = L/N \quad (8.10)$$

The van Deemter equation (for GC) identifies the fact that the plate height, H, of a column depends on a number of factors, including the linear velocity, $\bar{u}$, of the mobile phase, as follows:

$$H = A + \frac{B}{\bar{u}} + C\bar{u} \quad (8.11)$$

where the various terms are due to different mechanisms which each cause peak broadening, i.e. A represents eddies in the fluid flow around particles in a packed column (not present in open-column systems), B represents diffusion of the analyte in the direction of travel, and C represents various mechanisms involving the transfer of the analyte between the two phases and across the mobile phase.

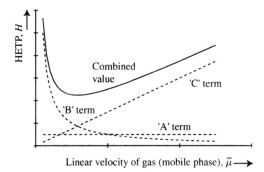

Figure 8.6 Linear velocity of the mobile phase (gas) ($\bar{u}$) as a function of the height of a theoretical plate (H) (the van Deemter relationship).

A plot of the van Deemter equation (Figure 8.6) shows that there is a linear velocity for which the plate height, H, is a minimum, i.e. it is possible to adjust the flow rate to maximize the efficiency of the separation.

There is a similar plot for HPLC but this has a flatter curve at higher velocities, thus allowing greater mobile phase velocities without significant loss of efficiency.

DQ 8.3

By which of the possible factors shown in the table below, will (i) the resolving power, R_P, and (ii) the peak width, W, change if the length, L, of the column were to be increased? Assume that the flow rate remains constant.

Factor	(i) Resolving power, R_P	(ii) Peak width, W
A	$R_P \propto \sqrt{L}$	$W \propto \sqrt{L}$
B	$R_P \propto L$	$W \propto 1/\sqrt{L}$
C	$R_P \propto L^2$	W is constant

Answer

We know that the resolving power is proportional to the square root of the number of plates, N (equation (8.9)), and that N is also proportional to the length of the column. Hence, assuming that all other factors are constant, the resolving power will increase as the square root of the length ((i), A). As the retention time increases in proportion to the length, we can use equation (8.8) to deduce that the peak width will also increase as the square root of the length ((ii), A).

From the above discussion question, we can derive the following equation:

$$\sigma = \frac{t}{\sqrt{N}} \propto \frac{L}{\sqrt{L}} \propto \sqrt{L} \qquad (8.12)$$

SAQ 8.2

The width, W, of a chromatographic peak is due to random diffusion of the analyte component as it passes through the column, and will increase as a function of the retention time, t_R, spent in the column.

The theory of the 'random walk' (see Section 13.5) suggests that W will be proportional to the following:

(a) $\sqrt{t_R}$ (True/False?);
(b) t_R (True/False?);
(c) t_R^2 (True/False?).

8.5 Separation Factor and Resolution

Are peaks resolved if they are separated?

The efficiency, plate number and capacity factor are all based on the column performance for a *single* analyte component. However, a main function of chromatography is to separate *different* components, and it is important to be able to quantify the success of any particular separation.

Figure 8.2 includes two components, P and Q, with different retention times. The *separation factor*, α, is defined as the ratio between the capacity factors for the two components P and Q, given by the following:

$$\alpha = \frac{k_Q}{k_P} \qquad (8.13)$$

where k_P and k_Q are the capacity factors of peaks Q and P, respectively. The separation factor is a function of the choice of stationary and mobile phases.

However, it is important to differentiate between 'separation' and 'resolution'. The ability to resolve two peaks depends on their distance apart in relation to their widths. If the widths of the peaks are greater than their separation, then they will overlap in the chromatogram and make resolution difficult or even impossible. The capacity factor does not give information about the widths of the peaks, and hence the separation factor *does not* give a measure of the resolution.

A measure of resolution can be obtained by taking the ratio of peak separation to the combined widths of the peaks, as follows:

$$R_S = 1.18\frac{(t_Q - t_P)}{(W_{HP} + W_{HQ})}; \quad R_S = 2\frac{(t_Q - t_P)}{(W_{BP} + W_{BQ})} \qquad (8.14)$$

For virtually complete resolution, we should have $R_S > 1.5$. However, lower resolutions can be tolerated, provided that allowance can be made for overlap when evaluating the measurement of areas.

By combining equations (8.3), (8.13) and (8.14) (and assuming peaks of equal width), it is possible to obtain the following expression:

$$R_S = \frac{\sqrt{N}}{4} \times \left(\frac{k_Q}{k_Q + 1}\right) \times \left(\frac{\alpha - 1}{\alpha}\right) \qquad (8.15)$$

Hence, the resolution of two peaks is a function of the following:

- the resolving power of the column, $\sqrt{N}$ (see equation (8.9));
- the capacity factor of the last peak, k_Q;
- the separation factor between the peaks, α.

DQ 8.4

Identify which of the experimental conditions (A–D), shown in the following table, are the main influences which affect the various performance factors (N, k_Q and α):

Factor			
N	k_Q	α	Experimental conditions
			(A) Temperature programming (GC) or solvent gradient (HPLC)
			(B) Flow rate
			(C) Column type
			(D) Column length

Answer

The plate number will depend on the column type and its length. It will also depend on the flow rate according to the van Deemter relationship (equation 8.11), and can be optimized by choice of the appropriate flow rate (N: B, C and D). The separation factor depends largely on the stationary and mobile phases (α: A and C). The capacity factor can be altered for different components by the use of gradient elution in HPLC or temperature programming in GC (k_Q: A).

The above discussion question identifies the fact that there are a number of options open to the analyst to change the conditions of the separation. It is therefore possible to change the experimental conditions to optimize the performance of the system according to different *criteria*, e.g. maximum resolution, minimum run time, etc.

Other factors will then set *limits* for the optimization of column design, with examples as follows:

- narrower columns (and smaller particles) increase the required *pressure* drop per unit length of the column, thus reaching a practical maximum;
- narrow columns use very small sample volumes, thus leading to problems of detector *sensitivity*;
- longer columns increase resolving power but will also increase the *analysis time* and the required *pressure* drop.

SAQ 8.3

For the chromatogram shown in Figure 8.2, calculate the capacity factors, separation factor, and resolution (using equation (8.14)) for peaks P and Q. From these values, deduce the plate number, *N*, of the column.

8.6 Detector Response

Does the choice of detector contribute to the 'experimental conditions', as well as the response characteristics of the measurement?

Chromatographic detectors have the same *general* characteristics as the generic signal-processing unit considered earlier in Section 3.2, i.e. responsivity, linearity, signal detectability, noise and drift. However, the individual detection mechanisms have their own *specific* characteristics, including selectivity, and these are discussed in specific sections for HPLC (Section 9.3) and GC (Section 10.3). The detectors can respond to different analytes and introduce different experimental conditions (Section 3.1) to the overall analysis.

Detectors are generally sensitive to either of the following:

- the *mass flow rate* of the analyte, e.g. in a mass-spectrometry detector or a flame-ionization detector;
- the *concentration* of the analyte, e.g. a UV-absorption detector or a thermal-conductivity detector (TCD).

A 'mass-flow-rate' detector records the *arrival* of the analyte molecules. The integrated response over *time* (peak area) will be proportional to the *total mass* of the analyte component and independent of flow rate.

A 'concentration' detector records the *presence* of the analyte, and the *integrated* signal will continue to increase for as long as the analyte is in the detector. The peak area is proportional to *concentration divided by flow rate*, with a slower flow rate giving a larger integrated signal.

DQ 8.5

Can I measure peak height instead of area?

Answer

*If the **width** of the peak does not change, then the height of the peak will also be proportional to its area. However, allowing for the possibility that the width of the peak may be dependent on the conditions of measurement and can vary from run to run, it is more reliable to integrate the total **area** in order to derive the analytical result, rather than rely simply on peak height.*

8.7 System Performance

What should I test to be sure that my chromatograph is working correctly?

The performance characteristics of a chromatographic instrument fall into the three main groups identified in Table 8.1, i.e. (i) Analytical Method Characteristics, (ii) System Characteristics, and (iii) Module Characteristics.

There is a 'top-down' hierarchy of dependence from the *method* characteristics (Section 1.5) to the *module* characteristics. For example, the ultimate *method* accuracy in the measurement of concentration will depend on the peak-area precision of the *system*, which in turn will depend, in part, on the precision of the injection *module*.

Table 8.1 gives the primary dependence of the various factors, but there can also be second order effects, e.g. a change of flow rate would affect the peak area for a detector that measures concentration.

In general, the various characteristics are evaluated as follows:

- *method* characteristics are recorded during Method Validation (MV);
- *system* characteristics are confirmed by using System Suitability Testing (SST) and Performance Verification (PV);
- *module* characteristics are confirmed by using Performance Qualification (PQ) and Operational Qualification (OQ) tests.

Jenke [2–4] has reviewed the method characteristics that are included within method validation procedures for chromatography.

The aim of the System Suitability Test as part of an analytical method is to confirm that the instrument *system* performance will satisfy the requirements of the *analytical method*. The most common tests for chromatographic *system suitability* usually include the following:

- the precision of retention times;
- the precision of peak areas;

Table 8.1 The three main performance characteristics classes for a chromatographic instrument

Method characteristics	Instrument system characteristics	Module characteristics
Accuracy	*Method* accuracy may depend on any of the following *instrument* characteristics	
Precision	**Retention-time precision**	Column (retention time)
		Flow-rate precision
		Solvent-proportioning precision (HPLC)
		Jacket-temperature precision (HPLC)
		Oven-temperature precision (GC)
	Peak-area precision	Autosampler precision
		Injection-volume precision
		Detector precision
		Column (bleed)
Bias	**Systematic factors**[a] Experimental conditions to which the method *may* be sensitive	Injection carry-over
		Injection-volume bias
		Autosampler bias
		Solvent proportioning
		Flow rate
		Jacket temperature
		Detector-specific conditions, e.g. wavelength accuracy and drift
Linearity	**Response linearity**	Injection linearity
		Detector linearity
Detection limit	**Signal-to-noise ratio**	Detector responsivity, noise and drift
		Column bleed
Selectivity (specificity)	**Resolution**	Column properties, e.g. efficiency, resolution, tailing and purity
	Choice of detector	Specificity of detector
	Detector selectivity	Wavelength accuracy and spectral bandwidth, e.g. for a UV detector

[a]For example, wavelength accuracy and spectral bandwidth for a UV detector.

- the peak resolution;
- the detection limit.

These will be confirmed by repeatedly running a standard sample under the defined operating conditions, and then making appropriate measurements on the resulting chromatogram.

Most chromatographic analyses employ a comparative method — hence the run-to-run repeatability (precision) is normally considered to be the most significant characteristic. However, a confirmation of absolute performance values (e.g. flow rate accuracy) is also important in order to ensure that the *conditions* of the experiment (Section 3.1) match those established during the method-validation process.

The *selectivity* (or *specificity*) of the method will depend on the following:

- the resolution of the column;
- the conditions of the separation (mobile-phase proportioning and temperature programming);
- the choice of detector;
- the conditions of detection (e.g. wavelength used in HPLC).

The *detection limit* and *linearity* for a method will depend on a number of issues related to system responsivity and system noise — these have been previously discussed in Sections 3.2.4 and 3.3.

The VAM guidelines [5] identify relevant characteristics and tests at both PQ and OQ levels for typical HPLC systems. Clearly, the performance of the column itself is a crucial factor in such tests. However, the OQ tests are strictly related [5] to the performance of the instrument *modules*, and do not therefore test the column, although they may use a standard column as part of the test procedure.

It is useful, however, to design the testing procedure so that a combined SST/PQ process satisfies both *system* and *module* objectives. For example, finding that the retention-time precision is within acceptable limits will confirm (SST) that the system, including the column, is performing satisfactorily, but it will also confirm (PQ) that those specific modules, which affect retention time, are also performing within their limits.

SAQ 8.4

Why are peak *areas* measured in chromatography, while peak (line) *heights* are generally measured in UV–visible spectrophotometry?

Note. Further questions that relate to the topics covered in this chapter can be found in SAQs 3.5, 11.4, 12.2 and 13.3, and DQs 11.2, 13.9, 14.8 and 15.1.

Summary

This chapter has quantified the structure of the chromatogram (for both GC and HPLC) by identifying parameters that are used to describe the important characteristics. It discussed the relationships between the descriptions of column efficiency, separation, resolution and peak shape.

Various options to optimize the performance of the chromatograph, by changing the experimental conditions, were related to their effects on the different performance parameters.

Finally, 'top-down' and 'bottom-up' relationships were identified between module performance, system performance, and the performance characteristics of the analytical method itself.

References

1. Dolan, J. W., *LC/GC*, **15**, 1018–1020 (1997).
2. Jenke, D. R., *Instrum. Sci. Technol.*, **25**, 345–359 (1997).
3. Jenke, D. R., *Instrum. Sci. Technol.*, **26**, 1–18 (1998).
4. Jenke, D. R., *Instrum. Sci. Technol.*, **26**, 19–35 (1998).
5. National Measurement System VAM Programme, *Guidance on Equipment Qualification of Analytical Instruments: HPLC*, 1998. [Copies available from VAM Helpdesk, LGC (Teddington) Ltd, Teddington, UK.]

Chapter 9

High Performance Liquid Chromatography

Learning Objectives

- To identify factors that affect the performance of individual HPLC modules.
- To relate the performance of the overall system to the performance of separate modules.
- To appreciate the difference between holistic and modular system testing, and the relationship to equipment validation and qualification.

9.1 Basic System

'HP' stands for 'high performance' in HPLC — could 'HP' also stand for 'high pressure'?

High performance liquid chromatography (HPLC) has become a rapidly developing standard separation technique, due to the following facts:

- it is suitable for those compounds which are not amenable to gas chromatography (GC) (i.e. non-volatile compounds and those which dissociate on heating);
- it has also proved to be an accurate quantitative analytical method.

An HPLC system is based on chromatographic separation in a 'column' with liquid solvent as the carrier medium. The generic characteristics of column separation have been described earlier in Chapter 8.

HPLC columns use very close packing with very small particle sizes in order to achieve sufficient column efficiency. These very small particle sizes produce a high resistance to fluid flow, and it is essential to use high-pressure pumps in order to obtain a sufficient flow of the mobile phase. The high *pressures* in HPLC are a consequence of the development of high-*performance* columns.

The basic system for HPLC (see Figure 9.1) requires the following components:

- a pumping system (pumps, mixer, etc.);
- a sample-injection system;
- a column;
- temperature control of the column (not always used);
- a detector system.

HPLC systems can be managed on a *modular* basis in which the user can connect different modules (e.g. column and detector) depending on the analytical requirements. This contrasts to GC systems that are assembled before delivery by the manufacturer, using 'modules' as ordered by the user.

Figure 9.1 shows two solvent supplies, which are mixed to provide the mobile phase. When using two, or more, solvent sources, it is possible to automatically change the proportions of individual solvents in the carrier medium during the run — this is known as *gradient elution*.

In practice, common systems may be described as follows:

- *isocratic*, where an unchanging solvent mixture is delivered;
- *binary* (as shown in Figure 9.1), where two solvents are mixed in a ratio which changes at a predetermined rate;
- *ternary/quaternary*, where mixtures of three or four solvents, respectively, are varying in a predetermined way.

Gradient HPLC is very useful for *developing* new methods as it gives the analyst an additional experimental parameter, namely the *solvent concentration*, which can be adjusted to obtain optimum separation. However, the *reproducibility* of a gradient method is not as good as an isocratic method because the retention times may be very sensitive to slight variations in the gradient-proportioning profile.

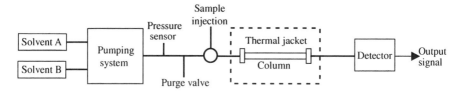

Figure 9.1 Schematic representation of a basic HPLC system.

The pumping system (see Section 9.2) is required to provide sufficient *pressure* to drive the fluid through the *resistance* of the column at a determined *flow rate*. The latter is determined by the physical speed of the pistons used in the pumping system. The pressure caused by the resistance of the column is called the 'back-pressure', and the pump must have sufficient power to maintain this pressure when providing the flow. For a column with a low resistance to flow, the back-pressure will be low, and vice versa.

The sample is 'injected' between the pump and the column. Although the earliest systems used a syringe injection similar to that used in GC systems, the high pressures in HPLC make it impracticable to develop an accurate and robust syringe injector. Most 'injection' systems now use a sampling loop system, in which a rotating valve allows a short length of tubing to be switched into the path of the solvent. The sample is introduced into this loop while it is out of the solvent flow, with 'injection' occuring by switching the loop into the solvent path.

The injection system is often automated by the use of an autosampler which draws up the sample from one of a set of vials in a carousel, and then passes it to the sampling loop. The volume injected may be determined, either by the volume of the loop directly, or for a *partial-fill* system, by the volume drawn up by the autosampler.

DQ 9.1

There is always some imprecision in the autosampler and injection system (e.g. 0.5% *RSD*). Which of the following statements are true:

(a) injection imprecision contributes directly to retention-time imprecision;

(b) injection imprecision contributes directly to peak-area imprecision;

(c) a comparison of the chromatogram of the test sample with that of a standard sample (external standard) can be used to compensate for the effect of injection imprecision;

(d) a comparison of the peak area of the test analyte with that of an added (spiked) known standard (with a different retention time) can be used to compensate for the effect of injection imprecision?

Answer

Injection imprecision affects the amount of sample introduced, and will therefore affect the amount of sample (peak area) detected (b, true). The retention time will not be affected (a, false).

Injection imprecision relates to the random differences in injection volume between runs, and hence will introduce uncertainties between the recording of different chromatograms (c, false). However, if the peak area of the test analyte can be compared with the peak area of a different *standard in the **same** chromatogram, then it is possible to monitor (and compensate for) the effect of injection-volume variability (d, true).*

The temperature of the column affects the separation process, and fluctuations in laboratory temperature may significantly reduce the repeatability of the process. The column can be enclosed in a *jacket* which is thermally controlled in order to maintain a constant temperature.

Finally, the eluent passes from the column to the detector (see Section 9.3). Some detectors monitor a certain *bulk property* of the solvent (e.g. refractive index) that may be affected by the presence of the analyte, while other detectors measure a property which is *specifically characteristic* of the analyte, e.g. its UV absorption or fluorescence.

DQ 9.2

The analyst has many operational choices when setting up an HPLC analysis. For each of the choices shown in the following table, identify whether they affect the experimental conditions of the analysis (Type I), the response (peak area) of the system (Type II), or both (refer to Section 3.1 for explanations of Types I and II).

Choices	Type I	Type II	Types I and II
Flow rate			
Solvent(s)			
Injection volume			
Column			
Detector			

Answer

*The flow rate, solvents and column all affect the **conditions** of the separation (Type I), but do not directly affect the response of the system to the analyte. The choice of injection volume will have a direct effect on the signal **response** (Type II). Detection is part of the experimental process in that the conditions of detection (e.g. wavelength of detector) can introduce selectivity between different analyte components (Type I). In addition, the sensitivity of the detector also determines the responsivity of the system (Type II).*

Various additional elements in the basic HPLC system include the following:

- a solvent degasser, which removes dissolved gas before the solvents enter the system — this prevents the gas from forming small bubbles at later stages in the process;

- a purge valve, which enables the pumping system to be flushed out when changing the solvents;
- a pulse damper, which absorbs the pressure pulses due to the reciprocating pumps;
- pressure sensors, which check that the pressure in the system does not become too high (indicating some blockage) or too low (indicating a possible leak);
- a guard column, introduced before the main column, which absorbs components that could contaminate the main column;
- filters, which remove impurities that might otherwise contaminate the column.

The unoccupied 'dead volume' in the system between the point of injection and the point of detection should be as small as possible in order to reduce dispersion of the sample, as this would result in peak broadening. Consequently, the piping should be kept as narrow and as short as possible and joints need to be made cleanly and without introducing any additional voids.

9.2 Pumping System

Why is there such a variety of different types and specifications of pumping systems?

The pumping system can be considered to include solvent-mixing and pulse-damping facilities in addition to the actual pump. In practice, such systems are often produced commercially as integrated units.

The objective of the system is to deliver the *correct composition of solvent mixture*, at a *pre-determined flow rate*, and with a *minimum of pulsation* in the flow rate.

Common *analytical* separations typically use flow rates in the range 0.1–10 ml/min. Preparative separations use larger sample volumes and wider-bore columns, requiring flow rates up to about 1000 ml/min. At the other extreme, microbore systems, with very narrow columns, operate with flow rates of 10–400 µl/min.

Some typical specifications for fluid flow could be as follows:

- *flow accuracy*: 1.00% or 0.05% of the maximum flow (whichever is the greater);
- *flow precision*: 0.1% or 0.005% of the maximum flow (whichever is the greater).

These would be measured under defined conditions, e.g. at 1 ml/min and 1000 psi, with water as the fluid.

Pumping systems must be capable of driving the fluid against high *back-pressures*, with typical systems operating up to pressures of the order of 6000 psi or even higher.

DQ 9.3

HPLC pumps often have safety limits at both high and low pressure levels, which automatically switch off the pump if the limits are reached. Give examples of circumstances that could cause the following:

(i) the high-pressure limit is reached;
(ii) the low-pressure limit is reached.

Answer

*If the column becomes blocked, the resistance to flow would be very high. The pump, in trying to maintain a constant flow rate, would push harder — the pressure would **increase**. The high-pressure limit would switch the pump off before the pressure became too high for the safety of the components.*

*If, however, the connections have not been made correctly and fluid was able to escape, then the resistance to flow could be very **low**. Without the low-pressure switch to turn the system off under these circumstances, the solvent could be forced to leak rapidly from the system.*

Stepper motors are normally used to drive the pumps. The coil windings of this type of electric motor are arranged so that the shaft rotates in very small *steps* as a result of the electrical pulses applied to the motor. Thus, the rotation speed of the motor is determined exactly by the *rate* at which the pulses are applied — the speed does not depend on the load that is being driven (provided that the motor has sufficient power). This can be contrasted with a normal electric motor whose speed is far less controllable and can be affected by the load that the motor is trying to drive.

9.2.1 Performance Tests

Flow-rate accuracy and precision can be measured without the use of a column by recording the actual volume of fluid delivered in a known time. Pumping the fluid through a capillary constriction allows this measurement to be made with the pump working against a realistic back-pressure. The performance of the pump will vary under different conditions of pressure and fluid viscosity, but will often be quoted using water under standard conditions, e.g.

Flow rate accuracy: $\pm 1\%$ of Control Setting at 1 ml/min and 1000 psi with water.

For gradient systems, the proportion of one solvent (e.g. B) in a mixture (of A and B) is an important performance factor. This can be measured by adding (to solvent 'B') a compound (e.g. acetone) that can be identified by the HPLC detector. The (detector) signal is then proportional to the percentage of solvent B in the mixture. A plot of signal against time gives the solvent ratio as a function of time, and allows the accuracy, precision and linearity of the proportioning system to be measured. It is also possible to record any short-term fluctuations (ripple) that the pumping system may induce in the solvent proportions.

9.2.2 Syringe Pump

The simplest pump system consists of a syringe-type piston, which drives the fluid through the system with a single 'one-off' stroke. The volume of the syringe limits the *total* fluid volume that can be delivered in each 'run'. In practice, this type of pump is limited to the provision of the very low flow rates suitable for narrow-bore columns.

DQ 9.4

A syringe pump has the following characteristics:

> Syringe volume: two × 10 ml microsyringes for isocratic and binary gradient operation;
>
> Flow rate: 1–2000 µl/min, at increments of 0.1 µl, with 4 nl stepper-resolution.

Calculate the following:

 (i) the frequency of the pump 'steps' when operating at 1 µl/min;
(ii) the maximum length of time for a run when operating at 400 µl/min.

Answer

 (i) At a rate of 1 µl/min, the system will deliver 1 µl in 1 min. Converting this to seconds, the rate becomes 1/60 = 0.0167 µl (= 16.7 nl) delivered in 1 s. We are told that one step of the piston provides 4 nl, and hence the number of steps in 1 s (the frequency) will be given by 16.7/4 ≈ 4.2 steps/s. At higher flow rates, the step frequency will increase proportionally.

 (ii) The total volume of the syringe, and hence the maximum volume that can be delivered, is 10 ml (= 10 000 µl). If 400 µl are being delivered every minute, this reserve of fluid can only last for 10 000/400 = 25 min.

Although the usefulness of the syringe pump is limited by its maximum volume, it does have the significant advantage that the flow rate does not suffer from the pulsations that can occur with reciprocating pistons.

9.2.3 Reciprocating Pumps

For the total volume required by most HPLC applications, a single-piston syringe pump is not a practical proposition, and it is necessary to use reciprocating-piston systems.

However, any reciprocating-pump system faces the difficulty of maintaining the flow when the piston is returning to refill the pump barrel with new fluid. This leads to *pulsations* in pressure and flow.

A *pulse damper* is capable, in the short term, of absorbing much of the excess (or deficit) *flow* due to the pressure pulsations. Its operation is equivalent to a flexible diaphragm which moves in or out in response to a change in flow, thereby absorbing the pressure fluctuations. However, a damper system will introduce a significant increase in the dead volume of the system.

There are many different commercial designs for pumping systems, and all manufacturers produce their own design and develop particular innovations. A common design uses two pistons, as illustrated in Figure 9.2. In this figure, piston A draws in the fluid and transfers it to piston B. The pistons then force the fluid to the output with a system of one-way valves to control the directions of flow. The one-way check valves operate by using a simple ball and socket arrangement, so that the fluid can flow easily one way, but, with reverse pressure, the ball sits firmly in the socket and thus blocks any reverse flow. The check valves are normally constructed from hard-wearing sapphire or stainless steel, but nevertheless are still subject to wear and will require replacement after various periods of use.

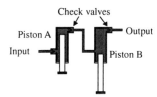

Figure 9.2 Schematic diagram of a reciprocating pump used in HPLC systems.

A standard design, capable of binary operation, uses a single motor to drive the twin pistons via cams. This basic unit can then be used, together with different pump 'heads', to produce pumping systems which are capable of different ranges of flow rates.

DQ 9.5

If pressure pulsations do not alter the **average** flow rate, why is it important to reduce pressure fluctuations as much as possible?

Answer

The detector may be sensitive to pressure changes and allow the pulses to 'appear' as interference in the chromatogram. The pressure pulsations can also have a gradual detrimental effect on the column packing.

Another practical problem to be overcome in a reciprocating pump is the need to draw the fluid from the external reservoir into the pump barrel. If this happens too quickly, the reduced pressure that is used to 'suck' the fluid through the valves can cause cavitation (voids created in the fluid) and will also allow any dissolved gas to form bubbles. This will affect the pump performance.

In order to reduce the possibility of gas bubble formation, it is often necessary to degas the solvents, by using either helium sparging or vacuum degassing. Care must be taken with the former method to ensure that selective evaporation of the constituents of the mobile phase into the helium flow does not occur, thereby altering the solvent proportions.

The pumping system must be designed to give good performance in fluid delivery, but it must also be capable of working for a long time without mechanical failure or excessive maintenance. The pistons use seals that are subject to wear, as well as chemical and biological attack, and they must be replaced periodically. For special applications, it is possible to obtain pumps made from titanium or chemically inert polymers. Biocompatible pump heads, that are automatically washed after each stroke, are also available.

The exact flow rate can be affected by compression of the fluid at different pressures. For many pump systems, the operator can enter the value of the solvent *compressibility* into the control system of the instrument, which then monitors the fluid pressure and automatically adjusts the piston drive to compensate for this compressibility.

SAQ 9.1

The description of a range of HPLC pumps includes the following specifications:

- *Flow-Rate Ranges*   Pump 'A'   *0.025–25 ml/min,*
                                 Pump 'B'   *0.05–50 ml/min,*
                                 Pump 'C'   *0.1–100 ml/min;*

- *Flow-Rate Accuracy*   ±*1.00% of selected flow rate or* ±*0.05% of maximum flow (whichever is the greater).*

  (i) For Pump 'B', calculate the flow-rate accuracy as a percentage of the *selected flow* across the whole range (e.g. use selected values, such as 0.05, 0.5, 5.0 and 50 ml/min).

  (ii) At which end of the range is the flow rate least accurate?

  (iii) Why would a manufacturer produce different pumps whose flow-rate ranges overlap to such a large extent?

9.2.4 *Gradient Systems*

For a binary system, it is common practice to use separate pumps for each of the solvents, and then to mix them *at high pressure after* the pumping stage. The composition of the mixture is controlled by the *flow rates* of the separate pumps.

However, it is not practical to have three/four separate pumps for ternary/quaternary systems, and the solvents are therefore usually mixed *at low pressure before* entering a single pump. *Proportioning valves*, which open sequentially for specific time periods, control the relative volume of each solvent that enters the pump, and hence control the composition of the mixture.

An important consideration for non-isocratic systems is the *delay volume*. This is the volume in the pumping system between the point at which the solvents mix and the exit point from the system. For low-pressure mixing, which occurs before the pump, the delay volume includes that of the pump. High-pressure mixing for binary systems occurs after the pumps and will have a smaller delay volume.

DQ 9.6

The specifications for composition of a quaternary pump are as follows:

Composition Accuracy: typically ± 1% from 5 to 95%, up to 5 ml/min.

What practical feature in the pumping system could be limiting the accuracy of composition below 5% and above 95%?

Hint — think carefully about how the proportions are determined.

Answer

*In a proportioning system, the valves will open and close for set periods to allow specific volumes of the different solvents to enter the pump. If one solvent is only present in a small proportion, then there will be a large differentiation between the periods for which the different valves are open. Small errors in the delivery of a small volume can easily cause large relative errors in the overall **proportion**.*

SAQ 9.2

In a gradient HPLC run, the pressure is seen to change even though the flow rate remains constant. Does this indicate a problem? If so, what is it?

9.3 Detectors

Can similar performance characteristics be used for different types of detector?

HPLC detectors either measure a bulk property of the eluent (e.g. refractive index) or a specific property of the analyte (e.g. UV absorbance). The performance characteristics include generic *response characteristics* (Section 3.2) which are appropriate to all detector systems, such as responsivity, linear dynamic range, noise and drift. In addition, there will be those characteristics which are specific to HPLC (e.g. cell volume and maximum pressure), plus those that are specific to the particular detection process (e.g. bandwidth).

The detection process occurs in a small flow cell, through which the eluent must pass. The effect of the *volume of this cell* on the measured signal is considered later in Section 15.2.

In the following sections, we will discuss the specific performances of four of the most common types of detector in current use. The use of a mass spectrometer as a detector is described in Chapter 12.

9.3.1 Variable-Wavelength UV Detector

Many analytes show absorption (particularly in the UV range), and consequently a spectrophotometric system (single- or double-beam) can be used as an HPLC detector.

Some 'fixed-wavelength' detectors use intense 'line' sources that give radiation at specific wavelengths (e.g. a low-pressure Hg lamp at 253.7 nm). However, a *variable*-wavelength UV detector uses a continuous source (e.g. a deuterium lamp) together with a monochromator to select desired wavelengths.

The absorption process takes place within a small cell, through which the mobile phase is continuously flowing. The choice of shape and volume for this *flow cell* depends on two main factors, as follows:

- the absorbance, and hence the detector sensitivity, will be proportional to the optical *pathlength* (assuming Beer's Law);
- a large cell *volume* will result in 'box-car' broadening of the lines (see Section 15.2).

Consequently, a common design for the flow cell has a cylindrical shape with the optical path along the axis of the cylinder. This maximizes the pathlength for a minimum volume.

The design of the cell should also seek to reduce the following:

- the amount of turbulence that occurs within the cell, as this can cause refraction of the light, thus giving apparent changes to the amount of light transmitted;
- the possibility that small 'bubbles' could be formed, thus giving apparent absorption peaks and/or noise.

DQ 9.7
What spectral bandwidth would be appropriate for a UV detector?

Answer

*Most analytes have a **broad** absorption in the UV range. We will see late in Section 15.5.1 that the spectral bandwidth can be up to about 10% of the analyte linewidth. In practice, most detectors have a spectral bandwidth of about 5–8 m, with a wavelength accuracy of ± 1 nm.*

An important practical issue is that HPLC solvents also absorb UV light at short wavelengths. This can have the effect of masking the true absorbance of the analyte, and therefore the use of high-quality solvents is required for short-wavelength measurements.

9.3.2 Diode-Array Detector

The diode-array detector (DAD) system uses a standard UV photometric arrangement, except that in this case the exit slit of the monochromator is replaced by a photodiode array (see Section 18.4), as shown in Figure 9.3.

The spectrum produced by the diffraction grating is spread along the length of the array. Each photodiode acts as a separate channel to record a different 'wavelength' in the spectrum, and transfers its signal into the data-processing section of the instrument. Each of the photodiodes records the light intensity falling on it at the same time as (i.e. in parallel with) the other photodiodes.

The spectrum can be 'scanned' by using computer software without the need to rotate the diffraction grating. There are no mechanical movements, and hence the problems of mechanical wear and backlash, which limit conventional scanning systems, are not present in this detector. Even during calibration, there is no need to make any fine adjustments to the position of the grating, as it is possible for the *software* to identify which diode element, n_x, corresponds to the calibration wavelength, λ_x. In addition, the wavelength repeatability of the system is excellent.

As the detector is an integral part of the dispersive system, the sample must be placed *before* the diffraction grating. Unfortunately, in this position the sample receives the full UV intensity of the source radiation, and there is therefore a possibility of fluorescence from the sample. Note that in the layout of a standard UV–visible spectrophotometer, the sample is placed *after* the monochromator

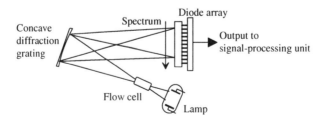

Figure 9.3 Schematic representation of a diode-array detector.

so that it is only exposed to the wavelength at which the absorbance is being measured.

A diode-array system has some very distinct advantages, as follows:

(i) The wavelength selection does not require any mechanical moving parts. This gives very good short- and long-term wavelength repeatability.

(ii) The measurement of all wavelengths of the spectrophotometer occurs simultaneously, thus giving a parallel multichannel detection. The full spectrum can be recorded very quickly in real time as the sample elutes, thus giving a *three-dimensional plot* of concentration against both *time* and *wavelength*.

(iii) The fact that the data for all wavelengths are recorded separately, but in parallel, allows sophisticated data-handling techniques to be used. If, for example, the analytical spectrum has a wavelength region of low absorbance, then it is possible to use a system of *internal referencing* which measures the light energy at a low-absorbance wavelength, at the same time as the measurement at the analytical wavelength. This monitors any possible variation in the light output from the light source itself, thereby compensating for the effect of source drift.

(iv) The computer system can independently convert each diode signal into an absorbance value, A_x. It is then possible to aggregate a number of diode signals together to provide fewer data points with an improved signal-to-noise ratio — a process known as *wavelength bunching*. We shall see later in Section 15.5.1 that it is possible to have wavelength bunching without sacrificing photometric linearity, provided that we aggregate the *absorbance* signals — this is possible with a DAD system. Note that opening the monochromator slit convolutes the *transmittance* signals, and will give a non-linear response unless the spectral bandwidth is significantly smaller than the signal linewidth (see Section 15.5.1).

The main disadvantage of the diode-array detector is the higher noise level which is inherent in the photodiode detection process. The responsivity (quantum efficiency) of the diode array is also temperature dependent, but this can be overcome by *temperature stabilization* of the detector chip.

DQ 9.8

A particular diode-array detector has facilities for the following:

(a) opening the entrance slit of the monochromator to give spectral bandwidths up to 10 nm;

(b) combining (bunching) the signals from adjacent diodes to cover equivalent spectral bandwidths.

Decide whether (a) and/or (b) could be used for each of the following circumstances:

 (i) measuring a broad spectral line, e.g. with a linewidth of 100 nm;
(ii) measuring a narrow spectral line, e.g. with a linewidth of 10 nm.

Answer

For the broad spectral line (i), it would be acceptable to open the slitwidth to at least 10% of the spectral linewidth without any significant loss of linearity (see Section 15.5.1). The diodes could also be bunched to the same extent (a and b).

For the narrow line (ii), a wide slit would cause a loss of linearity in the photometric response. However, it would still be possible to bunch the **absorbance** *(see Section 15.5.1) readings from the diodes (b).*

9.3.3 *Fluorescence Detector*

Using this method, the sample is excited by UV light (e.g. 200–400 nm) which causes the analyte to fluoresce at a longer wavelength (e.g. 300–600 nm). The light emitted due to fluorescence is recorded at 90° to the excitation direction. The wavelengths used for the *exciting* and *emitted* radiation beams are selected by separate monochromators, with typical spectral bandwidths set between about 5 and 50 nm. Each wavelength and spectral bandwidth can be set independently, according to the analytical method being used.

With this detector, it is possible to record both of the following:

- the excitation spectrum (monitoring at one emission wavelength);
- the emission spectra (keeping the excitation wavelength fixed).

DQ 9.9

A fluorescence detector uses a pulsed xenon lamp to provide the UV excitation radiation. The pulse rate, at which the lamp is fired, can be set by the operator at either 100 or 20 Hz.

Which of the following statements is/are likely to be true:

(a) the higher frequency can be used to increase the sensitivity of the detection;

(b) the higher frequency is used only for the shortest UV wavelengths;

(c) the lower frequency is used to reduce the noise in the output signal;

(d) the lower frequency will increase the lifetime of the lamp?

Answer

The 'trade-off' for the operation frequency of the xenon lamp is between the lamp lifetime and the average signal strength, and is not related to any specific radiation wavelengths (b, false). The higher frequency would be used to increase the average strength of the fluorescence signal for weak analytical signals (e.g. for trace analysis) (a, true). The lower frequency will increase the lamp lifetime (d, true), and would be used if the analytical signal was sufficiently strong. The reduced average signal strength at the lower frequency would also reduce the noise generated in the detector (see Section 15.4.2), but as the analytical signal would be reduced by a greater proportion, the S/N value would be lower (c, false).

9.3.4 Refractive-Index Detector

The refractive-index detector is a non-specific system that is useful for compounds that do not have a high UV absorptivity, such as polymers, sugars and organic acids. Changes in refractive index are very small and the system is sensitive to other physical perturbations such as *temperature* and *pressure* fluctuations. It is important, therefore, that the detector is thermostatically controlled in order to maintain a constant temperature, and that pressure pulsations are kept as small as possible.

DQ 9.10

Why is the refractive-index detector only used for isocratic measurements?

Answer

*The different solvents in a gradient measurement are likely to have different refractive indices, and as the proportion of solvents change, then the refractive index of the mixture will also change. This would give a **baseline shift** if a refractive-index detector was being used.*

SAQ 9.3

Under what conditions is it possible to have negative peaks in HPLC?

9.4 Performance Characteristics for HPLC Systems

If I can measure how well my instrument **system** *performs, why do I need to know how well each* **module** *performs?*

The specific processes of validating the performance of an HPLC system provides a very good illustration of the comparison between a *holistic* (whole system) approach and a *modular* approach to the verification of equipment performance.

Chapter 8 has examined the performance of a 'separation' instrument from a system perspective, and has developed the links to modular performance

(Section 8.7). Table 8.1 traces how the performance of individual *modules* may contribute to specific *system* characteristics.

In many cases, an uncertainty in *module* performance contributes directly to an *equivalent* uncertainty in the *method*, e.g. an imprecision in detector responsivity contributes directly to an imprecision in the calculated value of analyte concentration. However, because of the complex experimental processes occurring in analytical systems, an uncertainty in a *module* characteristic can often have an unexpected effect on the overall performance of the instrument.

DQ 9.11

Identify which of the **modules** (1–5) could affect each of the various system characteristics (A–E) given in the following table.

Module					System characteristic
(1)	(2)	(3)	(4)	(5)	
Pumping system	Injection system	Column	Thermal jacket	Detector	
					(A) Retention-time precision
					(B) Peak-area precision
					(C) Tailing factor
					(D) Noise
					(E) Resolution

Answer

*There are a number of obvious **first-order** connections that can be made, e.g. (1–A; 2–B; 3–A, C, E; 4–A, E; 5–B, D, E). There are also many **second-order** connections that can be made, particularly if the module is operating incorrectly, e.g. (1–D), if bubbles introduced through the pumping system can cause noise in the output signal, or (3–B) if column-bleed products interfere with the peak measurement.*

The process of confirming HPLC performance by using 'Equipment Qualification' (EQ) is a modular (bottom-up) approach (Section 4.12). The separate units (pumps, columns, detectors, etc.) may be obtained from different manufacturers and have separate lines of traceability for their individual EQs, in which case the user has the task of developing the common Performance Qualification (PQ) procedures. However, a single supplier may install individual modules as a combined system. In this case, it is probable that the supplier could develop a common Installation Qualification (IQ), Operational Qualification (OQ) and PQ procedure for the whole system.

The column is a replaceable item that is changed by the user to suit particular applications. It is *not used* in the formal EQ process until the PQ stage, where it is then only used to establish normal operating conditions for the system. The system as installed (IQ) and tested (OQ and PQ) by the supplier does not necessarily include any performance verification with the user's own column. Suppliers, in conducting an initial PQ test, would often use their own standard column. However, the supplier can sometimes co-operate with the user to develop specific test procedures appropriate to the specific requirements for quality management.

Some OQ tests can be performed on the module in isolation, e.g. the flow rate can be measured by recording the volume of fluid supplied in a given time. However, many 'modular' tests (e.g. proportioning accuracy) can be more conveniently performed by using a complete system under standard conditions, and can be carried out alongside the PQ tests.

Some of the modular characteristics identified in the OQ tests will contribute directly to the system performance which is tested by using the PQ procedure. Hence, a satisfactory 'pass' at PQ level could be used as confirmation of a 'pass' at OQ level.

The VAM Guidelines [1] identify the tests on individual modules that may be appropriate for OQ and PQ; these can be summarized as shown below.

- Pumping system (Section 9.2.1):
    fluid-flow-rate accuracy and precision;
    freedom from leaks;
    solvent proportioning — accuracy, precision, ripple and linearity.
- Sampling:
    precision, accuracy and linearity of injection volume (tested by direct measurement of sampled volume);
    autosampler thermostating.
- Jacket temperature:
    precision and accuracy.
- Detector performance (Section 9.3):
    responsivity — precision, accuracy and linearity;
    noise;
    drift;
    detector specific characteristics, e.g. wavelength calibration.

SAQ 9.4

A gradient elution in an HPLC system is found to give a *shifting baseline* when operating the detector at the short-wavelength end of the UV spectrum. However, the baseline shift does not occur when a constant (isocratic) solvent mixture is used or when the detection is made at longer wavelengths. Consider these symptoms carefully and suggest a possible cause for the shifting baseline, consistent with such observations.

SAQ 9.5

A colleague uses your chromatograph for a single run, but afterwards you find that the narrow peaks in your chromatograms appear to have been broadened. It does not appear to be due to a problem with the column. What could be the reason for this?

Note. Further questions that relate to the topics covered in this chapter can be found in SAQs 2.3, 3.2, 8.1, 14.1, 14.4, 17.2 and 18.3, and DQs 1.3 and 11.6.

Summary

This chapter has looked at the performances of the various modules that make up a standard HPLC system, and the effect that they have on the overall system performance. In particular, the performances of different pumping systems and various detectors were considered.

The HPLC system was used to illustrate the difference between holistic and modular performance testing, particularly in relation to the tests used for Equipment Qualification.

Reference

1. National Measurement System VAM Programme, *Guidance on Equipment Qualification of Analytical Instruments: HPLC*, 1998. [Copies available from VAM Helpdesk, LGC (Teddington) Ltd, Teddington, UK.]

Chapter 10

Gas Chromatography

Learning Objectives

- To identify factors that affect the performance of individual GC modules.
- To appreciate the problems in providing a controllable gas flow in GC.
- To interpret the specifications for various types of detector.

10.1 Basic System

Injection, separation and detection.

Gas chromatography (GC) is a very well established separation technique for the identification and quantification of materials that can be presented in a volatile state without decomposing. A carrier gas, typically He, N_2 or H_2, is the mobile phase. The stationary phase (solid or liquid) is held on a solid support material, which is either coated on the inside of a capillary column or held on particulate packing within a column. The general performance characteristics of chromatograph columns have already been described in Chapter 8.

The basic layout of a GC system, shown in Figure 10.1, includes the following components:

- a gas-flow control;
- a sample-injection system;
- a column;
- an oven;
- detector(s).

An essential feature of most GC methods is the ability to increase the temperature of the column during elution — known as *temperature programming*. This

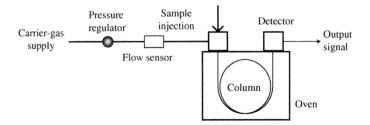

Figure 10.1 Schematic representation of a basic GC system.

enables analyte components with a wide range of affinities to be eluted within a reasonable time and with a reasonable peak structure and separation. The temperature programme starts at a low value, and is then raised at a predetermined *ramp rate*. At the low temperatures, those components with a low affinity will be eluted, and then the higher temperatures are used to limit the retention times of those components with a high affinity. The 'temperature–time' profile can be divided into several programmed sections to suit particular analyses.

Over the long history of GC, there has been considerable development in column design. Common systems now employ open capillary columns in which the solid support phase is coated on the inside wall of the capillary. Sizes vary from very narrow bore (0.10 mm internal diameter (ID)) to wide bore (0.75 mm ID) systems. The wider-bore capillary columns have sample-injection volume capacities that are comparable with conventional 2 mm ID packed columns.

10.1.1 Gas-Flow Control

The gas-flow rate through the column is determined by the pressure drop between the ends of the column. In manually controlled systems, the pressure is adjusted by using a regulator valve in the gas line from the supply. This sets the initial flow rate at the beginning of the analysis. However, as the oven temperature increases (with temperature programming), the viscosity of the gas will increase and the flow rate will fall.

DQ 10.1

The flow rate of a gas is difficult to measure mechanically because of the following:

(a) variations in viscosity and density between gases   (True/False?);
(b) variation of viscosity with temperature   (True/False?);
(c) high compressibility of gases   (True/False?);
(d) low gas densities   (True/False?).

Answer

Simple mechanical measures of gas flow are very susceptible to changes in gas properties for all *of the reasons given above.*

The above discussion question illustrates the problems inherent in measuring the flow rate of a gas. Historically, the main options for mechanical measurement had been the following:

- a spinning float that is forced higher up a conical tube by increased gas flows;
- a manometer to measure the pressure difference across a constriction in the flow.

Neither method was entirely satisfactory, and the best absolute measure of gas flow remained the 'soap-bubble' flowmeter at the output of the system, which measured the speed of a gas bubble along a calibrated tube.

It is now possible, by using stored data giving the properties of the gases, to develop accurate microprocessor-controlled gas flowmeters. Modern commercial GC systems offer an electronic control that is able to monitor the mass flow rate and continually adjust the pressure to produce the desired operating parameters. This is a significant improvement over manual systems, and it is possible to select an operation mode that keeps constant one of the parameters, e.g. pressure, mass flow rate, or average linear velocity. For example, the choice of the constant-flow-rate mode has specific advantages for those detectors (e.g. the thermal-conductivity detector (TCD)) whose integrated signal is dependent on gas flow rates.

SAQ 10.1

Identify **two** methods that could be used to ensure that 'late-eluting' GC components can be drawn out in a reasonable time without compressing the earlier part of the chromatogram.

10.1.2 Oven

The main performance specifications for an oven are as follows:

- *Temperature accuracy and precision* — the high sensitivity of retention times to temperature changes (a 1°C increase may cause a reduction of up to 5% in retention time) demand very accurate temperature control,

  e.g. $< 0.1°C$ below 150°C and $< 0.3°C$ below 300°C.

- *Ambient rejection* — the oven temperature must be unaffected by changes in the ambient laboratory temperature,

  where ambient rejection is defined as $\dfrac{\text{Change of oven temperature}}{\text{Change of external temperature}}$,

  e.g. < 0.02.

- *Heating ramp rate* — fast GC now uses a high ramp rate for the temperature programme (this requires an oven with a high-wattage heating element),

$$e.g. > 50°C/min.$$

- *Cool-down time* — a quick turn around between runs requires a short cool-down time,

$$e.g. \text{ cool-down time from 250 to } 50°C \text{ in 4 min.}$$

The maximum operating temperature is limited by the stability of the wall coatings of capillary columns; however, modern systems can now operate at temperatures of over 400°C.

DQ 10.2

Fans are used in GC ovens to:

(a) increase the heating ramp rate   (True/False?);
(b) increase the cool-down rate   (True/False?);
(c) improve temperature uniformity   (True/False?);
(d) reduce ambient rejection   (True/False?).

Answer

The accurate temperature control of the column is achieved by electronic-feedback control of the electrical power being supplied to the heater. It is therefore essential to have excellent thermal contact between the heating element and the column — this is achieved by the fan-assisted flow of air. The fan is mainly used to achieve excellent temperature uniformity (c) throughout the oven.

As can be seen from the above discussion question, the important features of a modern oven are the electronic control of temperature and a good thermal transfer between the heating element and the column. The ovens are also required to have separate heated sections for different elements of the system (i.e. the injection systems and various detectors), in addition to the main column section.

10.2 Injection Systems

Automated injection systems have reduced the effect of operator variability, but have not removed all uncertainty in the process.

There are many different injection methods and conditions, depending on the sample and column being used. The process of injection must be capable of the following:

- accepting the sample;
- vaporizing the sample, if required;

- transfering the sample (or a controlled proportion) to the column, without;
- decomposition of the analyte or;
- discrimination between different components of the analyte.

For vaporization, the sample is rapidly heated to an appropriate *injection temperature*, which must be sufficiently high to cause vaporization, *but not degradation*, of the sample. The sample is then swept into the column by the carrier gas. The vaporization conditions appropriate to one component of the analyte may not be appropriate to another — *discrimination* — with the result that some components are transferred more successfully to the column than others.

In some cases, it is possible to inject the cold sample directly into the end of the column and allow the vaporisation to occur due to the oven temperature itself — *cold on-column* injection.

For narrow-bore capillary columns, that can only accept small sample volumes and low flow rates, a split method of injection can be used to introduce a fixed *proportion* (e.g. 1/1000) of the original sample on to the column. However, any imprecision in the split ratio contributes directly to the imprecision in the system responsivity.

The success of the injector systems is measured by the extent to which the same amount of sample can be repeatedly injected. For off-column injection, uncertainties will occur due to imprecision in the following:

- transfer process to the column;
- split ratio;
- discrimination factor.

On-column injection gives a better *quantitative* measurement because there is no uncertainty due to a transfer process.

DQ 10.3

Which of the following is a disadvantage in using the split-injection method when making **trace** measurements:

(a) the split ratio is not sufficiently precise;
(b) a significant fraction of the analyte does not pass through the column;
(c) the noise will be greater?

Answer

*Using a split-injection method introduces an additional imprecision into the analysis, but this may not be significant when compared to the imprecision typically associated with **trace** analyses (a, false). However, the loss of a significant fraction of a small quantity of analyte does reduce the ability of the detector to record trace levels (b, true). The noise would not be directly affected by using the split-injection technique (c, false).*

The injection process for large numbers of samples can be automated with an autosampler. Each sample on a motorized carousel is drawn up in turn by using a syringe. Between injections, the autosampler must wash and rinse the injector system. If the wash cycle is not effective, there can be a *carry-over*, which contaminates subsequent samples.

DQ 10.4

Which of the following statements are correct:

(a) poor injection repeatability may give imprecision in peak areas;
(b) poor injection repeatability may give ghost peaks;
(c) carry-over may give imprecision in peak areas;
(d) carry-over may give ghost peaks?

Answer

Injection imprecision will contribute directly to peak-area imprecision, but will not lead to the development of ghost peaks.

*The effect of carry-over is less predictable, and will depend on the content of the previous sample. It is possible to introduce ghost peaks from **different** sample materials or increase the apparent analyte levels for **similar** samples. The effect of carry-over may appear as a systematic error (rather than an imprecision) in peak areas, thus making its presence difficult to detect.*

10.3 Detectors

Detection processes continue to develop in number and performance in spite of the maturity of the GC technique.

During its long history, a wide range of possible detectors has been developed for GC, either for specialist applications, or for more general routine use. The performance of GC detectors will include the following generic *operating* and *response* characteristics:

- maximum operating temperature;
- responsivity;
- linearity range;
- detectivity;
- selectivity;
- noise and noise filters (see Section 14.5.3).

In addition, there will be those characteristics which are *specific* to the particular detector being considered.

Table 10.1 Response characteristics of GC detectors

Detector	Responsivity	Detectivity	Selectivity	Dynamic range
Thermal-conductivity detector	10 μV/ppm	< 500 pg/ml (propane) < 1 ppm (nonane)		> 10^5 (± 5%)
Flame-ionization detector	> 0.01 C/gC[a]	< 4 × 10^{-12} gC/s (nonane) < 5 pgC/s (propane)		> 10^6 10^7 (± 10%)
Electron-capture detector		< 0.05 pg/s (lindane[b])	> 10^6 (Cl:hydrocarbon)	> 10^4
Nitrogen–phosphorus detector[c]		< 6 × 10^{-13} gN/s < 0.2 pgP/s		> 10^4 > 10^5

[a]Units are in coulombs per gram carbon.
[b]1,2,3,4,5,6-Hexachlorocyclohexane (γ-isomer).
[c]Modified form of the flame-ionization detector.

Table 10.1 gives the typical response characteristics of various detectors used in gas chromatography. The use of the mass spectrometer as a detector is discussed in detail in Chapter 12.

DQ 10.5

When comparing detectors, which of the following is the most useful characteristic:

(a) responsivity;
(b) detectivity;
(c) dynamic range?

Answer

The above characteristics all serve different purposes. Responsivity *gives the conversion of the analytical signal into another form, e.g. V, or C/s (= A). This is a useful parameter when matching the performance of the detector to the input of a separate amplifier. However, the range of analyte quantities that can be successfully measured by the detector is often limited by the noise produced in the detector itself, and the responsivity does not give any indication of such noise.*

However, detectivity, *which gives an indication of the minimum quantity that can be detected, and* dynamic range, *which gives the ratio of the maximum quantity to the minimum quantity, are both useful indicators. Nevertheless, such basic performance figures from the manufacturer should still be regarded with some caution, and the analyst should confirm, with the manufacturer or another user, the expected performance for their own particular application.*

An important characteristic of GC detectors is their *selectivity*. Detectors designed to measure specific analyte components must be able to do so in the presence of other common components. For example, the nitrogen–phosphorus detector (NPD) detects nitrogen and phosphorous but is insensitive to carbon, while similarly the flame-photometric detector (FPD) selectively measures sulfur and phosphorous, again in the presence of carbon, without any interference from the latter. The measure of selectivity is expressed as the ratio of the responsivities between the wanted and unwanted components, e.g. *the selectivity of the NPD detector for nitrogen is 2×10^4 gN/gC*.

A further characterization of the detectors divides them according to their dependence (or otherwise) on the mass flow rate — see Section 8.6.

The detector control unit, or software, frequently has the option of using a 'low-pass' noise filter (see Section 15.5.3) to smooth the noise in the signal.

DQ 10.6

What would be an appropriate filter time constant to use for a chromatogram with GC peaks which have 'full-width-at-half-height' (FWHH) values of 0.4 and 0.2 min?

Answer

The time constant should be as large as possible in order to reduce high-frequency noise, but not so large that any peak distortion will occur. In practice, the time constant should not be more that one tenth of the smallest linewidth. In this case, the value should be 0.02 min.

10.3.1 Thermal-Conductivity Detector

The thermal-conductivity detector (TCD) records the temperature of a heated filament in the gas flow. The presence of analyte changes the thermal conductivity of the gas, which in turn changes the rate at which heat is conducted away from the filament. When the conductivity decreases, the filament temperature will increase, thus leading to a change of resistance, which is measured electronically. This is the most universally used detector, and will respond to most analytes, although its sensitivity is relatively low.

DQ 10.7

The minimum detectable quantities for nonane for two detectors are given as follows:

    TCD   < 2 ppm;
    Flame-ionization detector (FID) 4×10^{-12} gC/s.

Explain why the detectivity of the same material (nonane) is expressed differently for the two detectors.

Answer

The TCD measures the conductivity of the gas/analyte mixture, which is a function of the concentration of the analyte in the gas. Hence, the detectivity is expressed in terms of the concentration.

The FID signal is proportional to the rate at which carbon compounds are ionized in the detector, and hence the detectivity is expressed in terms of the rate at which the atoms enter the detector, i.e. gC/s.

10.3.2 Flame-Ionization Detector

The flame-ionization detector (FID) is used extensively for hydrocarbons. With this detector, the eluent is introduced into an hydrogen/air flame, and a high voltage is applied between the flame jet (as one electrode) and a collector electrode. With no sample in the carrier gas, there is very little ionization to generate a current between the electrodes. However, carbon-containing molecules that burn in the flame will generate ions, which are then attracted to the collector electrode, thus causing a current to flow. The magnitude of the current is generally proportional to the carbon content of the analyte component.

Provided that the jet system is cleaned regularly and carefully, the FID is a robust and reliable detector, with a wide dynamic range.

DQ 10.8

Why is the FID such a popular detector?

Answer

Apart from its relative ease of maintenance, general reliability and good dynamic range, the FID has the advantage that it acts as a 'universal' detector, with a good sensitivity, for almost all compounds containing carbon.

The nitrogen–phosphorus detector (NPD) is a development of the flame-ionization detector, where in this case an alkali-metal-salt bead is placed in the flame. This has the effect of selectively increasing the sensitivity to nitrogen and phosphorus.

SAQ 10.2

A change in the mass flow rate will affect the total area of a GC analyte peak when using the following detectors:

(a) FID   (True/False?);
(b) TCD   (True/False?).

10.3.3 Flame-Photometric Detector

The flame-photometric detector (FPD) is used to quantify compounds containing sulfur or phosphorus by recording their photometric emission (Section 6.1) when the eluent enters an hydrogen/air flame. Optical filters select the wavelength appropriate to the specific atom to be detected.

The optimum flame temperature is different for different elements. A recent development has produced a Pulsed-Flame-Photometric Detector (PFPD) which repeatedly ramps the temperature of the flame in order to separate the *times* at which different components emit radiation. In this way, it is possible to further differentiate between the wanted and unwanted components in the signal, and thus increase the selectivity of the detector.

10.3.4 Electron-Capture Detector

The electron-capture detector (ECD) is selective for compounds that absorb electrons, e.g. halogens. A radioactive source emits β-radiation, which then ionizes some of the carrier gas (nitrogen or argon) to produce free electrons. A potential difference between two electrodes records the presence of the electrons as an ionization current. If an analyte component, which is capable of *capturing* the electrons, is eluted, then the measured current will fall in value.

DQ 10.9

Which of the following factors might affect the choice of the level of the standing ionization current due to the β-radiation in an electron-capture detector:

(a) with a **large** standing current, noise and drift in that standing current will mask small analyte-signal changes   (True/False?);

(b) a **small** standing current would be difficult to measure, and cause errors in trace measurements   (True/False?);

(c) with a **small** standing current, a large analyte signal will absorb a significant proportion of the current   (True/False?).

Answer

*Although a very small standing current increases the difficulty of measurement, the vital factor is the measurement of the **change** in the current due to the analyte — (b) is not a primary factor in this case. However, both factors (a) and (c) are important, and affect opposite ends of the dynamic range of the analyte signal. These dual limitations lead to a non-linear response and reduce the **dynamic range** of the detector. Compare this situation to that which occurs with the photodiode-array detector (PDA) in the discharge of each capacitor (see DQ 18.3).*

The ECD is a commonly used detector, and although limited by its non-linear response and poor dynamic range (see the above discussion question), its performance is steadily being improved by continual developments in the detection process.

SAQ 10.3

The description of a GC detector includes the following specification:

Signal Filtration   50, 100, 200, 500.

How would you interpret these figures? You should be able to work out what they mean even if you don't already know.

SAQ 10.4

A thermal-conductivity detector is specified with the following response characteristics:

Sensitivity: 8 µV/ppm (nonane);
Minimum detectable quantity: < 1.5 ppm (nonane);
Linearity: > 10^5.

(i) Explain the difference between 'sensitivity' and 'minimum detectable quantity'.
(ii) What assumptions or omissions have been made in these specifications?

Note. Further questions that relate to the topics covered in this chapter can be found in SAQs 3.3, 3.4, 8.1, 11.5 and 14.3, and DQs 1.9, 3.4, 8.1, 13.3 and 14.7.

Summary

This chapter has looked at the performances of the various modules that make up a standard GC system, with particular emphasis being placed on the performances of the injection system and the various detectors that are available.

GC detectors were used as good examples for examining the quantitative response characteristics of detectors in general, including responsivity, detectivity, selectivity and dynamic range.

Chapter 11

Capillary Electrophoresis

Learning Objectives

- To differentiate between the separation characteristics for CE and HPLC.
- To identify the factors that affect separation efficiency.
- To appreciate the limitations due to thermal dissipation.

11.1 Basic System

Capillary electrophoresis (CE) is a general name that describes a family of rapidly developing instrumental separation techniques [1, 2]. Each of these is a variant of the basic system in which a capillary, filled with an electrolyte buffer solution, links a source vial and a destination vial, as shown in Figure 11.1.

An electrical double layer is produced at the inner wall of the capillary, with the surface becoming negatively charged and attracting positive cations from the buffer. Some of these cations remain sufficiently mobile to be able to move *along* the wall. When a high voltage (up to 30 kV) is applied between the source and destination vials, the positive layer of cations is attracted towards the negative potential, and will drag the buffer solution towards the negative electrode with a velocity, v_{EOF} — this is known as *electro-osmotic* flow (EOF). An increase in pH or a decrease in the ionic strength of the buffer will affect the electrical double layer and increase v_{EOF}.

The basic analytical system is *capillary zone electrophoresis* (CZE), where the *separation* of analyte components occurs due to the differing mobilities of ions under the influence of the electric field. Positive ions will be attracted to the negative potential and will move *faster* than the buffer solution. Negative ions will be attracted to the positive potential, but because of the general flow of the

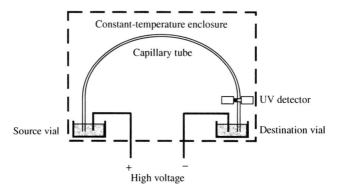

Figure 11.1 Schematic representation of a basic capillary electrophoresis system.

buffer solution, they will still move towards the negative potential, albeit at a *slower* rate. Neutral ions will move at the same rate, v_{EOF}, as the buffer solution. Unlike HPLC, there is no *stationary* phase in CZE.

The analyte components are recorded at different times by a detection system at some point on the column near to the destination vial. This produces an output similar to a chromatogram, but which, in this case, is called an *electropherogram*. The time between the injection of a component and its arrival at the detector is known as the *migration time*, t_m.

The 'injection' of a sample is achieved by momentarily moving the capillary input into a *sample* vial and then using one of a number of mechanisms to draw the sample into the capillary. The most common mechanisms include applying either (i) a pressure differential, or (ii) a voltage, between the sample and destination vials. The use of an electric potential to draw in the ions can have the advantage of 'selecting' only the *charged* ions for injection from the sample vial.

The capillaries are either made out of fused silica, which is particularly useful in that it is possible to measure UV absorbance directly through the capillary wall, or out of poly(ether ethes ketone) (PEEK), which is stronger but requires a separate detector cell.

The main advantages of CE are as follows:

- a greater separation efficiency (i.e. number of theoretical plates) over HPLC, by an order of magnitude;
- small sample volumes;
- a low volume of reagents is required;
- fast separation.

Capillary electrophoresis generally has higher (i.e. 'worse') detection limits than high performance liquid chromatography, but the sensitivity of CE is increasing rapidly as improvements continue to be made in detection processes.

11.2 Performance Characteristics

The commercial development of instrumental CE is still relatively new, with different configurations of CE systems being available from different manufacturers. This is due to the many variable parameters in CE systems, namely modes of operation, buffer and capillary characteristics, applied voltages, detection methods, etc. However, the overall objectives of all of the systems are similar and can be described by a common set of characteristics.

DQ 11.1

What would you consider to be the most important parameters that need to be checked as part of a system suitability check for a CE system?

Answer

A system suitability check should aim to measure those parameters that impact directly on the quality of the analytical results. As such, the CE parameters are likely to be very similar to those of the separation processes of GC and HPLC (see Table 8.1). Altria and Rudd [3] have discussed the various validation procedures for CE and suggest that the most important system suitability criteria for this technique may require measurement of the following:

- *resolution;*
- *migration time;*
- *injection precision;*
- *peak efficiency.*

A detection-limit (LOD) test may also be appropriate for trace measurements.

There are many similarities, and also some significant differences, between chromatograms and electropherograms, and it is useful to develop such comparisons by considering the characteristics previously developed in Chapter 8.

11.2.1 Migration Time

The migration time, t_m, of the analyte depends on the following factors:

- the electric field strength (E);
- the 'mobility', μ_{EOF}, of the electro-osmotic flow;
- the mobility of the analyte ions, μ_X, relative to the electro-osmotic flow, where mobility is defined as the velocity per unit electric field.

The observed velocity, v, of the analyte component 'X' is the sum of the velocities due to the actual migration of the component, v_X, and the electro-osmotic flow, v_{EOF}, as follows:

$$v = v_X + v_{EOF} = (\mu_X \times E) + (\mu_{EOF} \times E)$$

$$= (\mu_X + \mu_{EOF}) \times E = \mu_{eff} \times E \qquad (11.1)$$

where μ_{eff} is the effective (observed) mobility.

In addition, we can write the following:

$$v = \frac{L_A}{t_m} \qquad (11.2)$$

$$E = \frac{V}{L_{tot}} \qquad (11.3)$$

where V is the applied potential over the full length, L_{tot}, of the capillary, and L_A is the distance between injection and detection ($L_A < L_{tot}$).

Combination of equations (11.1), (11.2) and (11.3) gives the following expression:

$$t_m = \frac{L_A L_{tot}}{\mu_{eff} V} \qquad (11.4)$$

Imprecision in the observed migration times arises, in part, from the variations in electro-osmotic flow. The variation in t_{EOF} is greater than the variation in the equivalent value of the void time, t_M, in GC or HPLC.

DQ 11.2

As a means of compensating for variations in electro-osmotic flow, which of the following would it be better to use:

(a) an adjusted migration time ($t_m - t_{EOF}$);
(b) an effective mobility ($\mu + \mu_{EOF}$)?

(See Section 8.2 for comparison.)

Answer

*It is appropriate to use **adjusted retention** times in HPLC and GC, but by looking at equation (11.1) it can be seen that an **effective mobility**, μ_{eff}, is a more useful concept in CE.*

A neutral marker in the sample can be used to measure the electro-osmotic flow mobility, μ_{EOF}.

SAQ 11.1

A CZE electropherogram, using on-column detection, shows the EOF marker with a migration time of 2.0 min and an analyte peak at 4.5 min. If the applied voltage is 25 kV for a capillary length of 35 cm and an effective column length of 25 cm, calculate the mobility of the analyte.

11.2.2 *Effect of Temperature*

A consideration of the temperature effects is an important factor in capillary electrophoresis because of the following reasons:

- the supply of electrical power (voltage × current) into the capillary provides a direct heating effect;
- migration mobility increases significantly with temperature;
- an increase of temperature decreases the viscosity of the buffer, and increases the electro-osmotic flow.

The use of *narrow* capillaries reduces both the current flowing and, consequently, the amount of heat produced. The rise in temperature is also reduced by using some mechanism of forced cooling of the capillary — typically, a controlled air flow around the capillary.

Nevertheless, there will still be a temperature gradient across the capillary, with the highest temperature being in the middle. As a result, the migration mobilities of the ions in the centre of the capillary will be higher than those towards the capillary walls. This spread of mobilities contributes to a broadening of the analyte peak.

DQ 11.3

Which of the following happens if the applied voltage is increased beyond the normal maximum limit:

(a) the current increases in proportion to the applied voltage;

(b) the power dissipated increases in proportion to the square of the applied voltage;

(c) the migration time decreases in inverse proportion to the applied voltage?

Answer

All of these options would apply if the capillary acted as a simple electrical resistance with a constant effective mobility. However, as the power dissipation increases, the rise in temperature of the buffer increases the mobilities, and this in turn reduces the effective electrical resistance of the capillary. The falling resistance results in a further increase in both current and power, thus leading to yet higher temperatures and a potential 'runaway' situation.

From the above discussion question, it is clear that there is a maximum voltage beyond which the temperature gradient across the capillary increases rapidly (see Figure 11.2). If the temperature gradients become too large, then there will be a broadening of the peaks and an increased variability in the migration times.

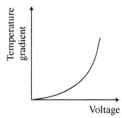
Voltage

Figure 11.2 Temperature gradient of the buffer as a function of the applied voltage in capillary electrophoresis.

SAQ 11.2

In a CZE measurement, the EOF mobility, $\mu_{EOF} = 3 \times 10^{-4}$ cm^2/(V s), changes by 2%. Calculate the resultant fractional changes in apparent migration times for the following:

(i) positively charged species with a mobility of 2×10^{-4} cm^2/(V s);
(ii) negatively charged species with a mobility of -2×10^{-4} cm^2/(V s).

11.2.3 *Efficiency*

The plate number, N, in CE defines the efficiency (see equation (8.7)), except that migration time in this case is used instead of retention time, as follows:

$$N = \left(\frac{t_m}{\sigma}\right)^2 \tag{11.5}$$

where σ is the standard deviation of the peak as a function of *time*.

Using $\sigma'_{tot} = v_X \times \sigma$, where σ'_{tot} is the standard deviation as a function of *distance*, and $L_A = v_X \times t_m$, the efficiency can also be expressed as follows:

$$N = \left(\frac{L_A}{\sigma'_{tot}}\right)^2 \tag{11.6}$$

Due to the fact that the electro-osmotic flow is 'driven' by the electrical force applied to the charged layer adjacent to the capillary walls, the velocity of the fluid is the same across the full width of capillary — see Figure 11.3(a). Thus, the inherent peak broadening is low in CE. This can be compared to the pressure-driven flow in HPLC, where the walls provide a frictional, retarding, force. The latter results in a velocity differential across the tube (see Figure 11.3(b)), and a spread of retention times that depend on whereabouts the analyte passes through the tube.

With such low peak broadening, it is possible to attain very high efficiencies (i.e. plate numbers) with CE.

However, various factors set minimum limits to the peak broadening, with each contributing an effective variance in the peak, as follows:

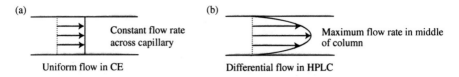

Figure 11.3 Comparison of the different flow rates in capillary electrophoresis and high performance liquid chromatography: (a) uniform flow in CE; (b) differential flow in HPLC.

- the length of the sample plug, l_S, in the capillary gives an effective variance, $\sigma_S'^2$;
- the longitudinal diffusion of the analyte gives an effective variance, $\sigma_{dif}'^2$;
- the temperature gradient across the capillary gives an effective variance, $\sigma_T'^2$;
- the detector cell length, l_D, gives an effective variance, $\sigma_D'^2$.

The combined effect of these factors can be obtained by adding their *effective* variances, as shown earlier in equation (2.7), thus giving a combined standard deviation *in distance*, σ_{tot}', as follows:

$$\sigma_{tot}' = \sqrt{\left(\sigma_S'^2 + \sigma_{dif}'^2 + \sigma_T'^2 + \sigma_D'^2\right)} \qquad (11.7)$$

DQ 11.4

Equation (11.7) uses an effective standard deviation, σ_S', to represent the broadening effect of the length of the sample plug, l_S. Which of the following is correct:

(a) $\sigma_S' = l_S/\sqrt{6}$;
(b) $\sigma_S' = l_S/\sqrt{3}$;
(c) $\sigma_S' = l_S/\sqrt{12}$;
(d) $\sigma_S' = l_S/\sqrt{2}$?

Answer

A plug of length l_S is equivalent to a rectangular distribution with limits of $\pm l_S/2$. This gives an equivalent standard deviation (Section 2.2.2) of $(l_S/2)/\sqrt{3} = l_S/\sqrt{12}$. The same relationship will apply in deriving the similar expression for the detector cell, i.e. $\sigma_D' = l_D/\sqrt{12}$.

Assuming that the system can be designed to reduce the other factors, the limiting factor in CE will be the longitudinal diffusion of the analyte. This is given by the following expression:

$$\sigma_{tot}'^2 \approx \sigma_{dif}'^2 = 2Dt_m \tag{11.8}$$

where D is the diffusion coefficient of the analyte.

Substituting for t_m from equation (11.4) then gives the following:

$$\sigma_{tot}'^2 \approx \frac{2DL_AL_{tot}}{\mu_{eff} \times V} \tag{11.9}$$

Equation (11.9) shows that, due to the reduced time spent by the sample in the capillary, an increase in the voltage *reduces* the peak broadening.

By combining equation (11.6) and (11.9), and using the approximation $L_{tot} \approx L_A = L$, we then obtain the following:

$$N \approx \frac{\mu_{eff} \times V}{2D} \tag{11.10}$$

Note, however, that the *maximum useful* voltage is limited by temperature effects — see DQ 11.3.

SAQ 11.3

A CE sample is injected by pressure for 4 s, and then produces a peak with a migration time of 9.0 min and a width at 'half-height' of 0.15 min. Is the injection period the main cause of peak broadening?

Estimate the effect on the peak width if the injection period were doubled to 8 s.

11.2.4 Resolution

The resolution between two peaks 'P' and 'Q' is given by an expression similar to equation (8.14), as follows:

$$R = 2\left[\frac{(t_P - t_Q)}{4(\sigma_P + \sigma_Q)}\right] \tag{11.11}$$

Note that the peak width at the base is 4σ.

If we then combine equation (11.11) with equation (11.5), we obtain the following expression:

$$R = \left(\frac{1}{4}\right)\left[\frac{(\mu_Q - \mu_P)}{\mu_{eff}}\right]\sqrt{N} \tag{11.12}$$

where μ_{eff} is the average effective mobility of the *two* components.

Then, by assuming that the peak broadening is due to longitudinal diffusion (from equation (11.10), the resolution is given by the following:

$$R = 0.177(\mu_Q - \mu_P)\sqrt{[V/(\mu_{eff}D)]} \tag{11.13}$$

DQ 11.5

The resolution in capillary electrophoresis can be improved by the following:

(a) increasing the length of the capillary   (True/False?);
(b) increasing the applied voltage   (True/False?).

Answer

Neither the efficiency (equation (11.10)) or the resolution (equation (11.13)) depend on the length of the capillary, but both of these, however, increase with the applied voltage.

From equation (11.10), it appears that the voltage should be increased as much as possible, but DQ. 11.3, however, shows that there is a maximum voltage limit, which is determined by temperature effects.

The analysis time can be reduced by reducing the length of the capillary, and from DQ 11.5 we can see that a reduction in the length of the capillary does not appear to effect the resolution. However, there is a lower limit to the useful length of the capillary due to the fact that a short capillary will offer low electrical resistance, thus giving a high current and an increased heating effect.

A careful balance therefore has to be struck between capillary length and diameter, heat dissipation, cooling, buffer properties and the applied voltage.

11.3 Detection

The most common detection process used in capillary electrophoresis is UV absorption or fluorescence, which is measured directly through the transparent silica wall of the capillary. The optical pathlength across a 'simple' capillary is very short ($\approx 50~\mu m$). However, this can be increased at the detection point by creating a bubble-detection cavity within the capillary.

Where analyte components do not show UV absorption, it is necessary to use *reverse absorption* by adding a UV-absorbing electrolyte to the buffer and making a recording when the absorbance drops due to the presence of the analyte. The absorbing analyte must be an 'ionic match' with the analyte being measured in order to avoid peak distortion.

The use of a mass spectrometer also presents an effective detection system as the fluid flow rate in CE is very low (see Section 12.2).

DQ 11.6

The **speed** of the analyte **component** through the UV detector in **HPLC** depends on the following:

(a) the mobile phase flow rate   (True/False?);

(b) the retention time of that particular component   (True/False?).

The **speed** of the analyte **component** through the UV detector in **CE** depends on the following:

(c) the applied voltage   (True/False?);

(d) the migration time of that particular component   (True/False?).

Answer

Clearly, flow rate and applied voltage are the respective factors that affect the rate of passage of the analyte through the detector for HPLC and CE (a and c, true), but what of the retention time and migration time?

*For HPLC, once the analyte has left the column, its speed will be dependent **only** on the flow rate of the mobile phase, and not on the retention time (b, false).*

However, for on-column detection in CE, speed differences between ions will still occur, with those having a shorter migration time continuing to travel faster through the detector (d, true).

From the above discussion question, the *time taken* by the analyte component to *pass through* the CE detector will be proportional to its migration time. Thus, the observed peak will be broadened by the same proportion. The effect of this with a detector that measures *concentration* (e.g. the UV detector) will be that the peak height will not be affected, while the peak width will increase in proportion to the migration time of the peak. The peak area will therefore also increase in proportion to this migration time.

DQ 11.7

For on-column detection in CE, it is useful to record the analytical signal as peak area divided by the migration time. Why is this so?

Answer

*The peak **area** is normally used as a measure of the amount of analyte being detected, but in CE this area is also proportional to the migration time. Normalizing the peak area by dividing by the migration time thus makes an allowance for the effect of variations in the migration time between runs.*

11.4 Alternative Modes of Operation

There are a number of important variants on the basic capillary zone electrophoresis (CZE) mode of operation. These all use the electro-osmotic flow to drive the solution through the capillary, but each mode employs significantly different separation mechanisms. Capillary zone electrophoresis, as discussed above, performs this separation on the basis of the ionic mobilities, i.e. the ratio of ion charge to ion size.

Micellar electrokinetic capillary chromatography (MEKC) has the important distinction of being capable of separating *neutral* atoms, which it does on the basis of their hydrophobicity. A detergent, added to the aqueous buffer, forms charged micelles that will travel through the capillary at a different rate to that of the buffer. The migration time of a particular analyte will depend on its partition ratio between being attached to the micelles or residing in the aqueous solution. Strongly hydrophilic molecules will remain in the buffer and travel at the speed of the buffer, while those that are strongly hydrophobic will travel with the micelles. The hydrophobicity of the particular molecule will determine at what *time* it appears at the detector.

Capillary electrochromatography (CEC) is a variant of CE that uses a stationary phase in the column to provide an additional chromatographic partitioning effect. In this technique, the fluid flow is still driven by the electro-osmotic flow and not by any external pumps.

Other variants of CE include capillary gel electrophoresis (CGE), which uses a gel-filled capillary to separate the molecules on the basis of their size, and chiral capillary electrophoresis (cCE), which is an effective alternative to ion chromatography for the separation of optically active molecules.

SAQ 11.4

The measurement of analyte quantity is carried out by recording different parameters in different instruments, including the following:

(i) the peak height for a spectrophotometric line;
(ii) the peak area for a chromatographic peak;
(iii) the peak area divided by the migration time for an electropherogram peak.

Explain the reasons why different methods are used for these three types of analytical techniques.

SAQ 11.5

What is the similarity in the ways in which the following detection processes produce a signal:

(i) the use of an electron-capture detector;
(ii) the use of a photodiode-array detector;
(iii) UV detection in CE for analytes which do not show UV absorption?

What effect does this have on the performance characteristics of the various detection methods?

Note. A further question that relates to the topics covered in this chapter can be found in SAQ 8.1.

Summary

This chapter has described the operation of capillary electrophoresis, and developed the comparisons with high performance liquid chromatography. It derived the particular advantages of electro-osmotic flow in terms of reduced linear diffusion, and identified the problem of susceptibility to temperature gradients across the capillary.

The use of on-column UV detection was described, together with the effect of migration time on the observed peak area. Finally, the various modes of a number of capillary electrophoresis variants (CEC, MEKC, etc.) were briefly introduced.

References

1. Baker, D. R., *Capillary Electrophoresis*, John Wiley & Sons, Chichester, UK, 1995.
2. Khaledi, M. G. (Ed.), *High Performance Capillary Electrophoresis*, John Wiley & Sons, New York, 1998.
3. Altria, K. D. and Rudd, D. R., *Chromatographia*, **41**, 325–331 (1995).

Chapter 12

Mass Spectrometry Systems

Learning Objectives

- To appreciate the choice of an appropriate mass-selection mechanism.
- To identify the relevant performance characteristics for hybrid systems.
- To appreciate the problems of interfacing with MS detectors.

12.1 Basic Systems

Mass spectrometry is itself a very important branch of analytical instrumentation [1], in which molecules (or atoms) are initially ionized (with a charge z), and then physically separated on the basis of their mass-to-charge ratio, m/z. As charge is quantized (multiples of the electronic charge, e), the charged molecule will exist only at specific, well-defined, values of m/z. This specificity makes mass spectrometry a very powerful tool for both the identification and quantification of analyte components.

The process of physical separation occurs, in either space or time, by using a variety of techniques based on the flight of the ions in electric and magnetic fields. The path of each ion will depend on the ratio (m/z) between its mass (m) and charge (z). By appropriate field design, it is possible to ensure that only ions with a specific value of m/z will arrive at a single detector.

SAQ 12.1

How can a mass spectrometer detector system with a mass range of 2000 be used in the analysis of proteins of mass 30 000?

There are several different system designs, which use different combinations of static (DC) electric and magnetic fields and radiofrequency (RF) fields. The mass 'spectrum' of detector output against m/z is normally achieved by varying the strengths of the fields. The main systems are described below:

(a) *Magnetic-sector* systems, where the ions are accelerated by an electric field into an orthogonal magnetic field, which causes them to follow an arc with a radius that depends on m/z. The arrival of an ion with a specific value of m/z at the detector depends on the strengths of the applied electric and magnetic fields. Magnetic-sector instruments are capable of very high resolution, particularly if a 'double-focus' system is employed, which uses the additional selectivity of an electric field to compensate for variations in ion velocities.

(b) *Quadrupole* systems, which use four parallel bars to produce a combined DC electric field and RF field. After initial acceleration, only ions with a specific m/z ratio can travel exactly down the central axis and reach the detector. Mass scanning can be achieved by changing the applied fields in fixed ratios.

(c) *Ion-trap* systems, which use an RF field to trap ions with specific m/z values into circular orbits. As the applied RF voltage is increased, the lighter ions leave the trap and are detected by an electron multiplier. In Fourier-transform systems, the presence of the trapped ions can be detected by the absorption of energy from the RF field at specific frequencies. This frequency 'spectrum' can then be transformed into a mass spectrum (cf. the FTIR spectroscopy system, Section 7.1)

(d) *Time-of-flight* systems, which measure the time-of-flight of the ions along a tube. The accelerated velocity of the ions depends both on m/z and the applied electric field. Hence their time-of-flight (TOF) across the system can be used as a measure of m/z.

DQ 12.1

Which factors determine the pressures that should be used inside the mass spectrometer?

Answer

In all systems, the ions must be free to follow an unimpeded flight. This means that the analysis chamber must be kept at a very low pressure ($\approx 10^{-6}$ torr (1.33×10^{-4} Pa)) in order to prevent collisions between ions and gas molecules. The introduction rate of the sample must also be kept very low to help maintain the low pressure.

The arrival of the ions is detected electronically by their charge, using an extremely sensitive electron multiplier (see the dynode system in the photomultiplier tube, Section 18.2).

The output of the detector provides a two-dimensional plot (mass spectrum) of signal intensity versus atomic mass (Da). The effect of multiply charged ions ($z > 1e$) is that they will appear at an apparent mass value that is an integer fraction of their true mass.

The processes used to ionize the molecules are discussed below in the sections describing the various hybrid systems. However, an important side effect of ionization is the possible fragmentation of the molecule. Fragmentation can be useful in identifying the structure of an unknown species through an analysis of the mass of the various fragments, but it does have the effect of reducing the signal intensity at the desired m/z ratio and producing other unwanted signals in other parts of the spectrum. Ionization processes that do not cause fragmentation are described as being 'soft'.

12.1.1 Performance Characteristics

Important instrument specifications include the following:

- Mass Range;
- Resolution;
- Scan Rate;
- Sensitivity;
- Ease of Maintenance.

The *mass range* gives the range of 'm/z' values that the instrument can record. This equates to the actual mass range of singly charged molecules, and is given in atomic mass units (amu) (Da). Molecules with higher masses can appear within the specified mass range provided that they have a sufficiently large charge.

The *resolution* is the ability to differentiate between different mass values, and is quantified by the following expression:

$$\text{Resolving Power, } R = \frac{m}{\Delta m} \tag{12.1}$$

where Δm is the smallest difference in the mass, m, that can be resolved. The 10% criterion for resolution in mass spectroscopy states that *two peaks (of equal height) are just resolved if the signal between the peaks drops to 10% of the peak height.*

The highest resolution ($\approx 10^5$) is achieved by double-focussed magnetic-sector systems, while single-focus and quadrupole systems will give resolutions of about 5000. Ion-trap and TOF systems have lower resolutions of about 1000, but these are being actively developed as useful detector systems. However, the resolution is not normally high for systems used as detectors, and the standard resolution is typically sufficient to differentiate between unit masses, i.e. 1 Da.

DQ 12.2

Calculate the resolving power necessary to achieve a minimum resolution of 0.8 amu if the mass range of the spectrometer is 1000 amu.

Answer

By using equation (12.1) for the resolving power, $m/\Delta m$, we obtain 1000/0.8 = 1250.

SAQ 12.2

Does the use of the term 'resolving power' convey the same meaning when applied to a monochromator and to a chromatograph as it does when applied to a mass spectrometer?

The *scan rate* is the rate at which the system can change the fields to measure different values of *m/z*. This is an important consideration when the mass spectrometer is used as a detector in chromatography, as it is necessary in this case to be able to record the mass spectrum within a shorter time (approximately a tenth) than it takes for a peak to elute. The reason for this is that it is important that the input signal to the spectrometer does not change too much during the scan period, otherwise it would give a biased distribution of intensity between the peaks of different masses.

The scan rate for magnetic-sector instruments is defined as the number of seconds required to record the spectrum across a mass range that increases by a factor of ten (namely seconds per decade), e.g. from 20 to 200.

The *sensitivity* of modern systems is extremely high, due to the high gain factors that are achievable with electron-multiplier systems, with the latter being similar to those use in photomultiplier tubes (see Section 18.2). There can be a trade-off between sensitivity and resolution, with very high resolution requiring fine focusing of the beam, which itself reduces the signal sensitivity.

The detection system can be operated in a number of different *modes* depending on the requirement of the analysis. These include the following:

- a spectral scan over selected mass ranges;
- recording the signal at a specific mass value — *selected-ion monitoring* (SIM);
- recording the signals (SIM) at a number of specific mass values — *peak hopping*;
- measuring the *total ion current* (TIC), thus giving a combined signal across all masses, e.g. when used as a chromatographic detector to produce an overall chromatogram.

When using selected-ion monitoring, the sensitivity for the specific ion being examined is very much increased.

12.2 Hybrid Coupling Systems

The mass spectrometer had been developed to act as a mass-selective detector (MSD) for a variety of hybrid instrumental systems, including the *tandem techniques* of gas chromatography–mass spectrometry (GC–MS), liquid chromatography–mass spectrometry (LC–MS), and capillary electrophoresis–mass spectrometry (CE–MS). In addition, *inductively coupled plasma* mass spectrometry (ICP-MS) is a powerful and rapidly developing technique for multi-element trace analysis, either in solution or in liquid samples.

The challenge for the use of the mass spectrometer as a detector in GC–MS, LC–MS and CE–MS is to be able to achieve the following:

(i) transfer the analyte (initially at near atmospheric pressure) to the spectrometer (at a very low pressure), together with a minimum amount of the mobile phase;

(ii) ionize the analyte molecules and introduce them into the analysis chamber;

(iii) scan the mass spectrum several times during the elution of each peak.

MS systems, by virtue of their need to maintain a very high vacuum in the analysis space, are only able to accept a low input flow rate of material. This initially hindered the development of hybrid systems, thus leading to a range of different types of interfaces for both GC–MS and LC–MS [2]. The current systems that are available reflect the tremendous advances that have been made in recent years in overcoming the practical problems appropriate to each technique.

For ICP–MS systems, the plasma already produces the analyte in an ionized state, and it is therefore only necessary to introduce the ions selectively into the mass spectrometer.

The main requirements that determine the type of MS system that is chosen to operate as a routine detector include the following:

• a fast scan rate;
• sufficient resolution;
• simplicity and reliability in operation;
• ease of maintenance and cleaning.

In practice, the quadrupole and ion-trap systems provide adequate performance, together with the benefit of being easy to operate, clean and maintain.

12.2.1 Gas Chromatography–Mass Spectrometry

The coupling problems in the GC–MS system have been considerably reduced through the use of capillary columns, which have low gas flow rates, i.e. of the order of 1 ml/min. It is now possible to introduce the gas from the narrow capillary columns directly into the mass-selective detector.

If necessary, it is possible to use a 'jet separator' to separate the analyte from the excess carrier gas. In this system, the gas is directed, as a high velocity jet, towards a small tubular input to the MS unit. A high proportion of the light carrier gas atoms (usually helium) diffuses out of the line of the jet and is then pumped away. Most of the heavier analyte molecules suffer little deviation and will therefore enter the spectrometer.

The ease with which the analyte is introduced into the MS unit allows the following standard methods of ionization to be used in GC–MS systems:

(a) *Electron ionization* (EI), which directs a beam of electrons across the input analyte flow, thus causing ionization of the molecules. The energy of the electrons is considerable higher than that required for ionization (< 15 eV), with the result that the excess energy causes a significant amount of fragmentation.

(b) *Chemical ionization* (CI), which uses a reactant gas (e.g. methane) which is itself ionized by an electron beam. The reactant ions then interact with the analyte molecules, thus transferring some of the charge of the latter. This is a relatively 'soft' method of ionization, and causes less fragmentation. In positive chemical ionization (PCI), the molecule acquires an extra proton, while for negative chemical ionization (NCI), which gives greater sensitivity, the analyte molecule must be a good electron absorber.

The coupling of a gas chromatograph with a mass spectrometer has proved to be a very successful combination. Important practical aspects of this tandem technique are described in detail in the text by Kitson *et al.* [3].

12.2.2 Liquid Chromatography–Mass Spectrometry

The difficulty in designing an interface between an HPLC instrument and a mass spectrometer is the high mass flow rate in the mobile phase of the chromatograph.

DQ 12.3

Calculate the vapour flow rate at atmospheric pressure (in ml/min) which is equivalent to a fluid flow rate of 1 ml/min for methanol with a density of 0.79 g/ml.

Compare the result obtained with a typical gas flow rate used in gas chromatography.

Answer

The mass of 1 ml of liquid is 0.79 g, which is equivalent to 0.79/32 mol. The volume of the vapour at atmospheric pressure and 25°C will be given by:

$$V = nRT/P = (0.79/32) \times 8.31 \times 298/(101 \times 10^3)$$

$$= 0.000\ 605\ m^3 = 605\ ml.$$

A flow rate of 605 ml/min is thus considerably greater than a typical GC flow rate of 1 ml/min.

The above discussion question demonstrates the large differences in the equivalent vapour/gas flow rates for HPLC and GC. Several methods have been tried in efforts to remove the bulk of the mobile phase before trying to introduce the eluent into the mass spectrometer. For example, in the *moving-belt interface* method, the eluent is sprayed on to a moving belt which carries the material through a solvent-evaporating chamber before entering the mass spectrometer. However, it is now possible to 'introduce' flow rates of the order of 1 ml/min directly into an ionization chamber by using spray ionization techniques.

With *spray techniques*, the fluid is sprayed from a tube held at a high voltage, into a chamber at atmospheric pressure, together with a drying gas which also functions as a *nebulizer* (e.g. nitrogen) — see Figure 12.1. The process of spraying creates charged droplets, which then continue to subdivide under their own electrostatic forces. The solvent evaporates rapidly from the small droplets into the drying gas, thus leaving the analyte ions. These ions are drawn electrostatically into the capillary entrance to the mass spectrometer. The spray technique produces very few ion fragments, although there can be some collision-induced fragmentation (CIF) of the ions before they are analysed.

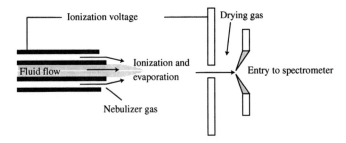

Figure 12.1 Schematic representation of the electrospray ionization technique used in LC–MS.

Chemical ionization (CI) is used for analytes that do not naturally form ions, and it is necessary in this case to add a reagent gas and heat the combined vapour, together with a corona discharge, to produce the ions. CI, being a more aggressive process, produces more fragmentation of the ions. Fragmentation can be useful in identifying the structure of the original ion, but it will reduce the signal in the SIM mode as the analyte signal is spread over several *m/z* values.

12.2.3 Capillary Electrophoresis–Mass Spectrometry

The principles involved in interfacing a capillary electrophoresis system to a mass spectrometer are very similar to those of LC–MS, except that the flow rates used in capillary electrophoresis are lower than those in HPLC.

When using the spray technique for ionization, it is common practice to add a coaxial sheath flow of liquid around the jet of the mobile phase which is being sprayed. This additional liquid film then makes electrical contact between the capillary and the electrode.

12.2.4 Inductively Coupled Plasma Mass Spectrometry

In this technique, the necessary ionization of the analyte is achieved within the plasma itself. In a typical configuration, the plasma is directed axially on to the apex of a cone with a small orifice at its peak — see Figure 12.2. The cooler outside sheath of the plasma is diverted away radially by the cone, and the core plasma can then pass into the MS analyser.

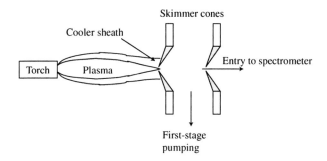

Figure 12.2 Schematic representation of the ICP-MS input stage.

DQ 12.4

A typical GC–MS system has a mass range of 1000 amu, while a typical ICP-MS system has a mass range of 250 amu. What is the reason for this difference?

Answer

The ICP system is used to identify and quantify elements. It is not therefore necessary to be able to record beyond atomic masses. The GC–MS system must be able to record a greater range of molecular masses.

ICP-MS suffers from a number of 'mass-spectral' interferences. These include the following:

- isobaric ions, e.g. tin and cadmium at 114 Da;
- polyatomic ions arising from the atomization process within the argon torch, with these mainly involving argon and oxygen;
- multiply charged ions.

However, ICP-MS is a very sensitive technique with very low detection limits for most elements, particularly when operated in the selected-ion monitoring (SIM) mode. Coupling this sensitivity with high scan rates across different elements thus makes ICP-MS a very efficient method for multi-element trace analysis.

SAQ 12.3

If I replace my flame-atomic absorption spectroscopy (F-AAS) instrument with an ICP-MS system, where the latter has a much wider dynamic range, which of the following statements will be true?

(a) ICP-MS will enable me to measure very low concentrations down to parts per thousand (ppt) levels.

(b) I won't need to dilute the samples that I currently use for AA spectroscopy.

(c) I can now measure trace elements in sea water, i.e. very low concentrations of elements in the presence of high concentrations of sodium.

SAQ 12.4

In ICP-MS, what is the value in recording the signals due to more than one isotope of a particular element?

Note. A further question that relates to the topics covered in this chapter can be found in SAQ 6.4.

Summary

This chapter has described the main elements of mass spectrometry, including the principal mass-separation mechanisms, and the use of a mass spectrometer as a detector in other analytical processes. It identified the performance characteristics appropriate to such a detection process, stressing the importance of fast scan rates to match the rate of production of data from the primary analytical procedure.

Each of the hybrid systems, namely GC–MS, LC–MS and CE–MS, was reviewed in respect of its particular characteristics, including the coupling between the two systems and the ionization processes involved.

Finally, the non-chromatographic technique of ICP-MS was discussed.

References

1. Barker, J., *Mass Spectrometry*, 2nd Edn, ACOL Series, John Wiley & Sons, Chichester, UK, 1999.
2. Abian, J., *J. Mass Spectrom.*, **34**, 157–168 (1999).
3. Kitson, F. G., Larsen, B. S. and McEwen, C. N., *Gas Chromatography and Mass Spectrometry*, Academic Press, London, 1996.

Chapter 13

Signals

Learning Objectives

- To estimate an appropriate sampling frequency in order to avoid aliasing.
- To identify important characteristics in the Fourier transforms of common types of signals.
- To estimate the bandwidth of frequency components in a Gaussian signal.
- To predict the effect of a low-pass filter on a noisy analytical signal.
- To be able to state the relationship between the time constant and bandwidth of a low-pass filter.

13.1 Introduction

What is the characteristic that is common to both of the following:

(a) the historic use of flags by ships for signalling;

(b) the high-pitched warbling tone that we would hear if we sent the same message by fax?

A *signal* requires a *change* in the value of some physical parameter. Examples of this include the following:

(i) the change, from red to green, in traffic lights which indicates when it is possible for a driver to proceed (with caution) to cross the road junction;

(ii) an alternating current (AC) signal at a desired frequency which may indicate that an aerial has 'picked-up' a local radio station;

(iii) a reduction in transmitted light intensity which indicates absorbance in an analytical sample;

(iv) a change in resistance in a thermal-conductivity detector which indicates the presence of an analyte in a GC gas stream;

 (v) a 'crackling' sound on the radio which may indicate that there is an electrical storm nearby;

(vi) a sea shell held close to the ear which may give a sound similar to that of waves breaking on a distant shore.

We can see from these examples that the *physical parameter* may take a variety of forms — voltage, light intensity, resistance, sound intensity, colour, etc. The list is endless, and also includes the different 'signal' flags used by ships.

In normal usage, the word *signal* also implies something about its *significance* to any observer, i.e. it conveys *information*. However, it is necessary to be selective about information — we are all bombarded by advertisements that try to provide us with information that we do not necessarily want!

The information conveyed in the first four examples (i–iv) given above is certainly *wanted* by the observer. In example (v), the 'crackle' on the radio also conveys information — it tells us about the storm — but is this information *wanted*, especially when it *interferes* with the enjoyment of the radio programme?

In contrast, the sound in the sea shell in example (vi) may appear to be 'information' (about a distant sea-shore), but this is, in fact, only random *noise* which is amplified by the shell so that it becomes audible.

An important conclusion is that it is only the *sender* or the *recipient* that can decide if the *signal change* carries *wanted information* or whether it is interference or noise. The instrument system that carries these signals can not identify any *purpose* for the change in the signal. Electronic circuitry can not, by itself, separate an unwanted signal from a wanted signal — they all appear as simple changes in a physical parameter.

In this present book, we will normally use the term 'signal' in its common usage to *imply* a transfer of information. However, where there is an issue about information content we will define our use of 'signal' more precisely in that particular context, e.g. to differentiate between an *analytical* signal and a *noise* signal. We will return to this point in Chapter 15 which discusses noise and interference.

13.2 Types of Signals

My age is an analogue quantity, but being 'over 40' is definitely digital!

There are two main types of signals, namely 'analogue' and 'coded' (a *digital* signal is a special case of a coded signal).

In *analogue signals*, the information is conveyed by the *magnitude* of the changing variable, e.g. the analogue output from an instrument may be capable of carrying a signal that can vary between 0 and 10 V. An analogue variable can take any possible value within its range.

There are many ways in which the information can be *encoded*, e.g. ships' flags, traffic lights, the Morse code, etc. However, it is necessary that both the sender and recipient *agree* on the code that is being used, e.g. 'red' means stop at traffic lights.

Analogue signals have the advantage of a continuous spread of possible values. Contrast this with the four (only) signal values used in the UK traffic-light system, i.e. red, green, amber, and red plus amber.

Coded signals have a specific advantage in that they are less susceptible to error. For example, if the intensity of the light in the red traffic light drops by 10%, it will continue to convey exactly the same information — it still means *stop*. In contrast, an equivalent drop of 10% in an analogue signal would represent an actual *error* of 10%.

DQ 13.1

How many different combinations of red, amber and green traffic lights are possible if at least one light must be 'on'?

Answer

We can represent an 'on' lamp by a '1' and an 'off' lamp by a '0'. Each lamp has two states, and thus the number of possible combinations is $2^3 = 8$, including the '000' state with them all 'off'. The answer is therefore 7 (8 − 1) possible states. In the UK, we only use four of these, i.e. (red/amber/green equals) 100 — stop, 110 — prepare to go, 001 — go, and 010 — stop if safe to do so.

13.2.1 Digital Signals

Digitization is the modern method of encoding signals in such a way that it is possible to keep the advantages of a coded signal while still being able to obtain good resolution of the analogue values.

A digital value is an agreed code of '0s' and '1s' which can be used to convey a series of number values. A 'byte' is a series of eight digits, each of which is called a 'bit'. Successive bits represent values which increase by powers of two, from the least significant bit (LSB) with a value of 1, to the most significant bit (MSB) with a value of 128. Each bit of a digital number identifies whether (1) or not (0) that particular 'power of 2' contributes towards the total equivalent analogue value. For example, the digital number 0100 1101 is equivalent to the analogue value 77:

	0	1	0	0	1	1	0	1	
	×	×	×	×	×	×	×	×	
MSB	128	64	32	16	8	4	2	1	LSB
	⇓	⇓	⇓	⇓	⇓	⇓	⇓	⇓	

$$0 + 64 + 0 + 0 + 8 + 4 + 0 + 1 = 77$$

Modern computer systems now operate with more 'bits' in each digital number. For example, '32-bit technology' is capable of processing 32 bit numbers directly instead of 8 bits.

DQ 13.2

Calculate the dynamic ranges (Section 3.2.4) of the analogue values that can be represented by the following:

(i) an 8-bit digital number;
(ii) a 16-bit digital number;
(iii) a 32-bit digital number.

Answer

The smallest (non-zero) analogue value recorded by all of the numbers is equivalent to the LSB, i.e. 1.

The largest value for the 8-bit number is $128 + 64 + 32 + 16 + 8 + 4 + 2 + 1 = 255$ ($2^8 - 1$). Thus, the dynamic range in this case (largest/smallest) is 255.

For the 16-bit number, the largest analogue value is $2^{16} - 1 \approx 6.56 \times 10^4$, while for the 32-bit number this value is $2^{32} - 1 \approx 4.29 \times 10^9$. Each of these values is then also equal to the equivalent dynamic range.

Computers can still process (e.g. add, multiply, etc.) signal values with dynamic ranges greater that those given in the above discussion question, but each operation will require more processing steps, and consequently takes more time.

A critical step in a computerized instrumentation system is the *conversion* of the (real-world) analogue signal to a digital signal. This conversion is performed by an electronic circuit called an analogue-to-digital converter (ADC).

DQ 13.3

A flame-ionization detector has a dynamic range of 5×10^6. Estimate which of the following number of bits would be used in the output of the ADC needed for this application:

(a) 12;
(b) 24;
(c) 32.

Answer

*A similar calculation to that used in DQ 13.2 would suggest that a 24-bit converter is required (dynamic range ≈ 1.7 × 10⁷). However, this would not be the best (electronic) solution. In practice, a 12-bit ADC is more likely to be used, in conjunction with **amplifiers** and **attenuators** which are used to bring the extreme analogue values back into an acceptable input range for the dynamic range of the 12-bit converter.*

SAQ 13.1

Which of the following is the most difficult process:

(a) converting a digital number into an analogue number;
(b) converting an analogue number into a digital number?

13.2.2 Sampling of Analogue Signals

An analogue signal is continuously variable in *time* as well as in magnitude. Before it can be digitized, the initial signal must be represented as a succession of analogue values — this process of representation is called *sampling*.

The frequency with which a signal is *sampled* is an important consideration. Figure 13.1 shows a wave 'A', of frequency, f_A, being sampled at a low relative frequency, $f_S (= 5/4 × f_A)$. It can be seen that it is possible to 'join the dots' by using a sine wave 'B' that has a frequency, f_B, lower than that of the original wave ($f_B < f_A$). In processing the signal, the computer will *only* have the sampled values to use, and would interpret the signal incorrectly as the low-frequency wave 'B' rather than the correct high-frequency wave 'A'. This effect is called *aliasing* and can be a serious problem, especially when unwanted signals

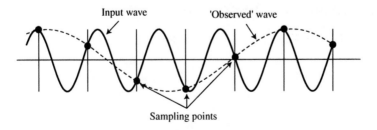

Figure 13.1 Illustration of aliasing in signal processing.

(e.g. noise) at high frequency are aliased in the computer at similar frequencies to those of the signal being measured.

In order to prevent aliasing, the sampling frequency should be *at least twice* that of the highest frequency component (see Section 13.4) in the signal, i.e. $f_S > 2 \times f_{A,max}$. This is known as the 'Nyquist criterion'.

DQ 13.4
Which of the following statements is/are true:

(a) high-frequency sampling is required for maximum dynamic range;
(b) high-frequency sampling is required for signals that have a 'complex' shape;
(c) the extra information obtained from sampling at a higher frequency is always useful;
(d) the only problem with sampling at a lower frequency could be a loss of high-frequency components?

Answer

*(a) The dynamic range (Section 3.2.4) is calculated along the **magnitude** axis of the signal, while sampling is along the **time** axis — they are* not *related.*

(b) The complex shape of a signal implies high-frequency components, and as such requires *high-frequency sampling.*

(c) Sampling at a frequency considerably higher than the frequency components in the analytical signal provides no *additional information about the signal.*

(d) This is definitely not true. *A high-frequency component sampled at a low frequency does not disappear, but will give a spurious signal at a lower frequency which may interfere with the analytical signal — this is the process known as* aliasing.

13.3 Harmonic (Sine or Cosine) Waves

Why is a sine (or cosine) wave the 'simplest' waveform?

It is important to realize that we live in an *analogue* world. The analytical signals that we are trying to measure will have analogue values, e.g. the concentration of a solution could, *in principle*, be measured to many decimal places, e.g. 3.034 803... g/l.

If we are to understand the performance of instruments, it is vital that we understand the structure of the information that the instrument is recording. We

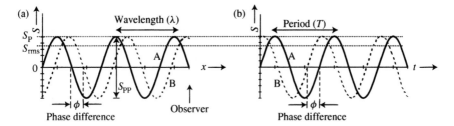

Figure 13.2 Illustration of the harmonic sine wave, showing its variation with (a) distance and (b) time.

will start with the simplest form of analogue signal, i.e. the harmonic sine wave, as shown in Figure 13.2.

A sine wave is a *periodic* analogue wave (or signal) — it repeats itself regularly with the *same shape*. The *magnitude* of the wave, S (Figure 13.2), is the value of this changing parameter at a given position, x, and time, t, and can be described by the following functions:

$$S = S_P \times \sin\left[2\pi(t/T - x/\lambda) + \varphi\right] \tag{13.1}$$

or

$$S = S_P \times \sin\left[2\pi(ft - kx) + \varphi\right] \tag{13.2}$$

where S_P is the *peak amplitude* (maximum magnitude), T is the *period* (time between points of equal phase in successive cycles), f is the *frequency* (number of cycles per second (Hz), where $f = 1/T$), λ is the *wavelength* (distance between points of equal phase in successive cycles), k is the magnitude of the *inverse wavelength wave vector* (number of wavelengths in one metre, where $k = 1/\lambda$), and φ is the *phase* (position within the cycle).

The magnitude of the wave can be measured as one of the following:

- the peak amplitude, S_P;
- the peak-to-peak amplitude, $S_{PP}(= 2 \times S_P)$;
- the root-mean-square amplitude, $S_{rms}(= 0.707 \times S_P)$.

The waves 'A' and 'B' shown in Figure 13.2 are identical, except for a difference in phase. The phase of the wave is measured as the fraction of the cycle that has been completed at that particular instant. By considering one cycle as the completion of a single rotation (i.e. 360°), the intermediate positions are measured as a phase angle between 0° and 360°. The phase angle can also be measured in radians, i.e. between 0 and 2π.

DQ 13.5

How many **independent** parameters are required to fully describe a given sine wave?

Answer

*The **shape** of all sine (harmonic) waves is already defined. It is then only necessary to define wavelength, frequency, amplitude and phase. If the velocity is already known (e.g. for an electromagnetic wave in free space), the frequency and wavelength are related (see equation (13.4) below), and it would then only be necessary to define one of the latter two parameters.*

The above discussion question shows that the sine wave is the simplest of waves. It is possible to completely describe the time dependence of the sine wave by using only three parameters, namely the amplitude, frequency and phase.

The difference between a sine wave and a cosine wave is only in the phase. A cosine wave 'leads' the sine wave by a phase difference of 90° or $\pi/2$ rad, as follows:

$$\cos\theta = \sin(\theta + \pi/2) \tag{13.3}$$

The *velocity*, v, of the wave is the distance that a particular point (or phase) on the wave will travel in one second. It can be calculated from the time, T, that it takes for the wavelength, λ, of one wave to pass an observer (see Figure 13.2), as follows:

$$v = \lambda/T = f\lambda \tag{13.4}$$

A *standing wave* is created when two identical waves are travelling in opposite directions. A guitar string, for example, can be considered as having two equal waves travelling backwards and forwards along the string in opposite directions, thus giving a combined oscillation that does not move along the string.

In instrumentation electronics, it is the behaviour of the signal *in time* which is usually most important. The behaviour with respect to distance is only important for particular situations, and usually only for waves of very high frequency (e.g. coaxial transmission lines).

DQ 13.6

Wavenumber (for mid-IR radiation) is calculated by taking the inverse of the wavelength in cm. Calculate the ranges in (i) wavelength (nm), and (ii) frequency, equivalent to the range 400–4000 cm^{-1}.

Answer

For $\sigma = 400$ cm^{-1}, $\lambda = 1/400 = 2.5 \times 10^{-3}$ cm $= 25\,000$ nm, while for $\sigma = 4000$ cm^{-1}, $\lambda = 1/4000 = 2.5 \times 10^{-4}$ cm $= 2500$ nm.

The velocity of the IR wave is equal to the speed of light ($v = 3.0 \times 10^8$ ms^{-1}). Therefore, for $\sigma = 400$ cm^{-1}, $f = v/\lambda = (3.0 \times 10^8)/(25\,000 \times 10^{-9}) = 1.2 \times 10^{13}$ Hz, while for $\sigma = 4000$ cm^{-1}, $f = v/\lambda = (3.0 \times 10^8)/(2500 \times 10^{-9}) = 1.2 \times 10^{14}$ Hz.

13.3.1 Addition of Sine Waves

We encounter many real situations where two or more sine (or cosine) waves (signals) combine together. The principle of *linear combination* (*Principle of Superposition*) can be stated as follows:

The magnitudes of all waves present at the given instant (t), *and at the given point* (x), *will add algebraically to give the total magnitude at that instant and at that point.*

We will illustrate this by adding the following sequence of sine waves: $S_1 = S_0 \sin(\omega t)$, $S_3 = (S_0/3) \sin(3\omega t)$, $S_5 = (S_0/5) \sin(5\omega t)$, $S_7 = (S_0/7) \sin(7\omega t)$, etc., where $\omega = 2\pi f$.

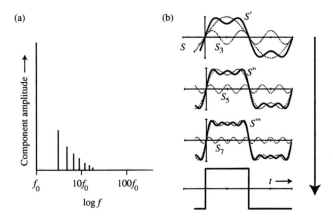

Figure 13.3 Synthesis of a square wave: (a) sine-wave components as a function of frequency; (b) combination of sine waves.

Referring to Figure 13.3, the addition of the first two waves, S_1 and S_3, is achieved along the time axis by adding their magnitudes at each point along this axis. In some places, both signals are positive and they give an enhanced signal, but elsewhere, one is positive and the other negative, thus resulting in a reduced net magnitude. The combination is shown as a new wave, S'.

Adding the next wave, S_5, gives a new intermediate wave, S'', in which we can see that the rising part of the wave has been made steeper, and the peak has been made flatter. If we continue to add further components, the top of the wave will become increasingly flat, and the rising and falling edges will approach vertical. An 'infinite' series of odd (3, 5, 7, 9, etc.) harmonics will eventually generate a *square* wave.

Figure 13.3(a) shows the sine-wave components (as a function of log f) which are added together, while Figure 13.3(b) shows the combination of the sine waves. The complete wave can be represented by the following equation:

$$S = S_0[\sin(\omega t) + (1/3)\sin(3\omega t) + (1/5)\sin(5\omega t) + (1/7)\sin(7\omega t) + \cdots\cdots]$$

(13.5)

DQ 13.7

What frequency ranges (high or low) of wave components are most important in giving a wave the following features:

 (i) **steep** rising and falling edges, e.g. in a square wave;
(ii) **sharp** corners, e.g. in a triangular wave?

Answer

 (i) *If you look at the construction shown in Figure 13.3, you can see that the steep edges for a square wave do not appear until the **high**-frequency components are included.*

(ii) *Similarly, any characteristic that requires a sudden change of slope will also require **high**-frequency components.*

In general, all complex waveforms will require components at high frequency.

SAQ 13.2

What waveform is produced when two sounds of similar amplitudes, but with frequencies of 1000 and 1002 Hz, are sounded together?

What will the result sound like?

13.4 Frequency Components

Why are the characteristics of the simple sine wave so important when 'real' waves always look far more complicated?

The previous section has shown how it is possible to *synthesize* a square wave by adding together a suitable mixture of sine waves of different frequencies and components. Hence, we know that, for a square wave at least, it would be possible to carry out the *reverse* process where the wave can be *analysed* into its separate *frequency components*, each of which is a pure sine wave.

The next important question asks if it also possible to analyse *any* wave into sine-wave frequency components in the same way as for the square wave?

DQ 13.8

Why does someone's voice often sound 'different' when you hear it over the telephone?

Answer

The telephone system has a limited 'bandwidth' and attenuates frequency components above a few kHz. The voice waveform will be distorted due

to a loss of some of the frequency components. It will still be recognizable
due to the low-frequency components, but will 'sound' different due to
the loss of high frequencies.

13.4.1 Fourier Transformation

It was the French mathematician, Jean Baptiste Joseph, Baron de Fourier, who
showed that it *was indeed possible* to analyse *any practical wave (*or *signal)*
into the sum of its harmonic (sine and cosine) wave-frequency components. The
mathematical procedure for performing the analysis is now called a *Fourier
transformation*, with the resultant set of frequency components being known as
the *Fourier transform* of the initial signal.

In the Fourier transformation, the information content of the wave is encoded
into the magnitudes of the parameters (amplitude, frequency and phase) which
describe each of the frequency components. These components can be drawn on
a plot of amplitude vs. frequency, as shown in Figure 13.4.

You will see in other texts that the formal mathematical presentation of the
Fourier transform includes both positive and negative frequency components. The
negative frequency is equivalent to a wave travelling in the opposite direction,
and is required to establish 'standing waves' (Section 13.3) in the time domain.
As the negative frequency components contain little *additional* information that
is significant for this present text, they have not been included in the diagrams.

In addition, it is not possible to show 'phase' on a two-dimensional plot (except
for a phase difference of 180° which would appear as a *negative* amplitude).
However, the omission of phase is not a limitation to understanding the concepts
involved.

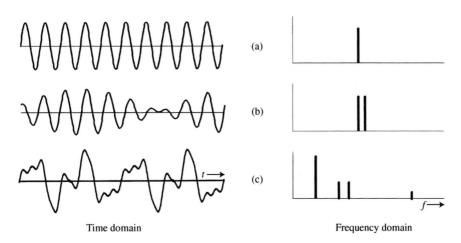

Time domain   Frequency domain

Figure 13.4 Fourier transformations of: (a) a simple sine wave; (b) the product of two
sine waves; (c) a complex signal.

The plot of the wave against time is called the 'time domain', while the corresponding plot of the frequency components is known as the 'frequency domain'. The Fourier transformation converts a signal in the time domain into its components in the frequency domain. The reverse process is known as an *inverse Fourier transformation*.

Figure 13.4(a) shows the simple sine wave and its Fourier transform, and Figure 13.4(b) shows the simple addition of two sine waves of slightly different frequencies. The envelope of the amplitude of the combined wave is sometimes called a 'beat' pattern, and can be heard when two sound waves have slightly different frequencies (e.g. when two aeroplane engines are working at slightly different speeds). Figure 13.4(c) shows a more complex, but still periodic, signal.

13.4.2 Periodic and Non-Periodic Waves (Signals)

The purity of the sine wave (see Figure 13.4(a)) is evident from the simplicity of its Fourier transform. An important characteristic of any *periodic* signal is that its Fourier transform has *discrete* components, i.e. the components occur separately at well-defined frequencies.

Fourier showed that it was also possible to identify frequency components for *non-periodic* waves, i.e. signals which do not repeat themselves. However, the Fourier transforms for non-periodic signals do not have discrete frequency components, but instead they show a continuous *distribution* of components, e.g. the Fourier transform of a Gaussian line has a spread of components whose amplitudes also plot out a Gaussian shape, as shown in Figure 13.5.

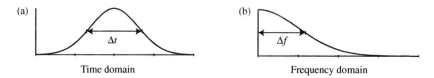

Figure 13.5 Fourier transformation of a Gaussian line shown for (a) the time domain and (b) the frequency domain.

When a periodic wave is limited in time, Δt, then the spread of components will cover a range of frequencies, Δf, which is given by the following relationship:

$$\Delta f = 0.44/\Delta t \tag{13.6}$$

where Δt and Δf are the widths at 'half-height'.

A narrow signal 'in time' requires a wider bandwidth of frequency components.

DQ 13.9

A chromatograph peak has a width of 0.1 min. Estimate the bandwidth of the signal components in the Fourier transform of this signal.

Answer

By using equation (13.6), for $\Delta t = 6$ s, we find that $\Delta f = 0.44/6 = 0.07$ Hz. The profile of frequency components that describe the chromatographic peak in the frequency domain will be given by a Gaussian curve (see Figure 13.5) with a bandwidth of 0.07 Hz.

SAQ 13.3

Figure 8.2 shows two chromatogram peaks, P and Q.

(i) Sketch the frequency components (Fourier transform) required to describe peak Q, including appropriate values on the frequency axis.

(ii) In a similary way, sketch the frequency components required to describe peak P.

(iii) How would the different positions in time be recorded in the Fourier transforms of the two peaks?

13.5 Combining Random Signal Components

Is it time to go for a 'random walk'?

The addition of components can be performed *graphically* by using vectors. The length of the vector, S_i, is proportional to the amplitude of the wave, with the direction angle, ϕ_i, being the phase angle. When we add together two waves of amplitudes, S_1 and S_2, and phases, ϕ_1 and ϕ_2, respectively, we get a combined wave of amplitude S_C, and phase ϕ_C, as shown in Figure 13.6.

If 'n' waves, each of amplitude, S_0, and with *the same phase (coherent)*, are combined, then they will add constructively to give a total amplitude $S_{\text{coherent},n}$, as follows:

$$S_{\text{coherent},n} = nS_0 \tag{13.7}$$

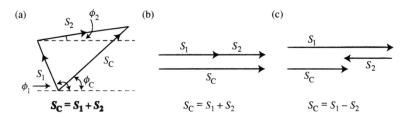

Figure 13.6 Combination of signal components: (a) vector representation (vectors shown in bold); (b) addition ($S_1 + S_2$); (c) subtraction ($S_1 - S_2$).

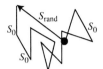

Figure 13.7 Combination of incoherent signal components ('random-walk' addition).

However, if they are in *random phase (incoherent)*, the addition is essentially that of the *random walk* — a series of steps of equal length, but in totally unrelated directions. Figure 13.7 illustrates such an addition.

Intuitively, we can imagine that (statistically) the most likely value for the sum of n steps of length S_0 in random directions will be as follows:

- less than $n \times S_0$;
- more than 0;
- a function of 'n'.

It can be shown that, for a large number of steps, the most likely value of $S_{\text{random},n}$ is given by the following expression:

$$S_{\text{rand},n} = \sqrt{n} \times S_0 \qquad (13.8)$$

This is an important result for the process of signal-averaging.

DQ 13.10

I take 100 steps of equal length, L, but with the direction of each step being independent of the previous step (**a random walk!**). What is the most likely distance that I will travel?

Answer

*Each step is similar to a random signal component, and the solution of this problem uses the same mathematics as developed above. The most likely **distance** travelled will be $\sqrt{100} \times L = 10$ L, although it is impossible to predict the **direction**.*

13.5.1 Signal-to-Noise Ratio for Replicate Measurements

The use of replicate measurements to selectively increase the analytical signal in comparison to a random error is introduced later in Section 14.5.5.

13.6 Analogue Filters

What do electronic filters 'filter'?

A filter is a signal-processing unit (Section 3.2.1) which displays the following properties:

(i) The output signal is in the same *form* as the input signal, for example:

- an electronic filter normally has a voltage signal at both input and output;
- an optical filter has optical signals at both input and output.

(ii) The maximum *gain* (or responsivity) is not normally greater than unity.

(iii) The gain is a function of the *frequency* or the *wavelength* of the signal.

The frequency responses of four common types of filters are shown in Figure 13.8. Each plot records the gain of the filter measured for sine wave components of different frequencies.

Figure 13.8(a) shows a filter for which the gain is 1 for components at low frequencies and which then falls to 0 for high frequencies. The effect of this is that a low-frequency component will pass through the filter without any reduction in amplitude, while a high-frequency component would be totally absorbed. This is called a 'low-pass' filter. The converse of this, i.e. a 'high-pass' filter, is shown in Figure 13.8(b).

Figure 13.8(c) shows a 'bandpass' filter where a band of frequency components can pass through the filter with both the high and low frequencies being absorbed, while Figure 13.8(d) shows a 'notch' filter, which can be used to absorb a specific frequency, f_0.

The gain (or responsivity) of a filter does not suddenly change between 0 and 1 at a precise frequency. We can define 'cut-off' points at high, f_H, and low, f_L, frequencies as being where the power gain (or optical intensity) of the filter drops to half of its maximum value. For an electronic filter, this is equivalent to a drop in the voltage gain to $\sqrt{0.5}(= 0.7)$ of the maximum value.

The *bandwidth*, Δf, of a filter defines the range of frequencies that can pass through the filter.

For the *bandpass* filter, the bandwidth is given by $\Delta f = f_H - f_L$.

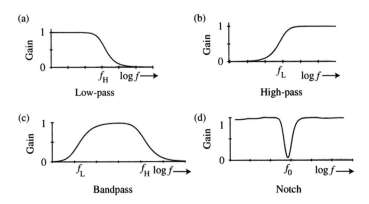

Figure 13.8 Plots of filter gain versus frequency for different filters: (a) low-pass; (b) high-pass; (c) bandpass; (d) notch.

In the case of the *low-pass* filter, the low-frequency cut-off is zero ($f_L = 0$) and hence $\Delta f = f_H$.

Figure 13.9(a) shows a chromatographic peak plus 'white noise', while Figure 13.9(b) shows the frequency components of the signal plus the noise (see Section 14.4).

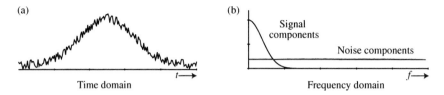

Figure 13.9 A chromatographic peak, including 'white noise', shown for (a) the time domain and (b) the frequency domain.

DQ 13.11

The signal plus noise shown in Figure 13.9(a) passes through a low-pass filter with a high-frequency cut-off, f_H.

(i) Draw a frequency diagram to show the frequencies appearing at the output of the filter.

(ii) What effect would this filter have on the appearance of the peak?

Answer

The effect of the filter must be calculated for each frequency component separately, by multiplying the input amplitude by the gain of the filter at that frequency, which gives the components shown in Figure 13.10(b). The result is that high-frequency noise components are absorbed, which gives the peak a smoother appearance — as shown in Figure 13.10(a).

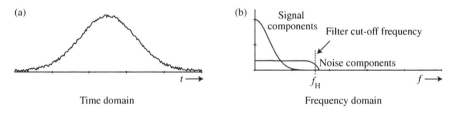

Figure 13.10 The signal shown in Figure 13.9 after passing through a low-pass filter to remove high-frequency noise components, shown for (a) the time domain and (b) the frequency domain.

A low-pass filter can be used to remove unwanted high-frequency (short-term) noise from a signal, by reducing the bandwidth of the system.

13.6.1 Signal-to-Noise Ratio for Filtered White Noise

White noise has signal components at all frequencies (see Section 14.4). These components are of *equal amplitude* but they are in *random phases* with respect to each other. In order to find the total noise, we must first find out which range of noise-frequency components will be recorded by the instrument.

Any frequency component (signal or noise) within the *instrument bandwidth*, Δf (see DQ 13.11), will be processed by the instrument. Outside the bandwidth range, all frequency components will yield a (near) zero response.

If the bandwidth, Δf, for a given system were to be increased, then there would be an increased noise contribution from the increased number of frequency components. If we assume that the noise amplitude for a *unit* bandwidth (of 1 Hz) is $S_{N,1}$, we can now consider how much the noise will *increase* if the bandwidth *increases* to Δf.

Within a range, Δf (Hz), the number, n, of different frequency 'components' (each of unit bandwidth) will be given by the following:

$$n = \Delta f / 1 = \Delta f \tag{13.9}$$

These noise components will add randomly as shown in equation (13.8), giving a total noise amplitude, $S_{N,\Delta f}$, for a bandwidth, Δf, as follows:

$$S_{N,\Delta f} = \sqrt{n} \times S_{N,1} = S_{N,1} \times \sqrt{(\Delta f)} \tag{13.10}$$

where $S_{N,1}$ is the amplitude of the noise for a bandwidth of 1 Hz.

Therefore, the total white noise increases as the *square root* of the signal *bandwidth* of the instrument.

The signal-to-noise ratio (S/N) for a bandwidth, Δf, is given by the following:

$$S/N = \frac{S_A}{S_{N,\Delta f}} = \frac{S_A}{S_{N,1}} \times \frac{1}{\sqrt{(\Delta f)}} \tag{13.11}$$

Provided that a reduction in the instrument bandwidth does not also attenuate the components of the signal, S_A, S/N will be proportional to $1/\sqrt{(\Delta f)}$.

13.6.2 Time Constant of a Low-Pass Filter

The representation of the low-pass filter in the *frequency* domain shows that it attenuates high frequencies. How can we represent its performance in the *time* domain?

Consider a step-change signal (as shown in Figure 13.11(a)) as being the input to a low-pass filter. The filter will absorb the *high-frequency* components, and

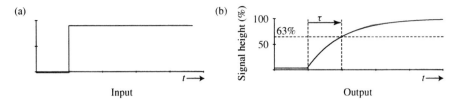

Figure 13.11 The effect of a low-pass filter on a step-change signal: (a) input; (b) output.

the steep rise of the step will be lost. The output from the filter will be as shown in Figure 13.11(b), which can be described by the following function:

$$V = V_0[1 - \exp(-t/\tau)] \tag{13.12}$$

The speed of response is characterized by the 'time constant', τ, where the latter is the time taken for the signal to reach 63% of its final height.

If the filter has a higher cut-off frequency ($f_H = \Delta f$), then it will also take a shorter time (smaller τ) to respond to a step input. This relationship is defined by the following equation:

$$f_H \tau = 1/(2\pi) \tag{13.13}$$

This relationship has a specific significance in the reduction of white noise by the use of electronic filters.

DQ 13.12

An electronic filter in the output from a (flame) atomic absorption spectrometer has a default time constant of 0.1 s. An analytical signal has a S/N value of 5 with the default setting. Estimate S/N if the time-constant of the filter is increased to 2 s.

Answer

*As the time-constant of the filter has **increased** by a factor of 20, the bandwidth will be **reduced** by the same factor. We can then see from equation (13.11) that S/N will increase by a factor of $\sqrt{20}(= 4.47)$, to give a value of 22.4.*

13.7 Optical Filters

Is a Michelson interferometer a form of optical filter?

The transmittance, T_λ, of an optical filter is wavelength-dependent, thus giving a spectral transmittance profile that is characteristic of that particular filter. Optical filters have been designed for a range of different purposes.

Wavelength selection can be achieved with common filters, which use the selective absorption of coloured glass or dyed gelatine. However, these have a relatively broad bandwidth, e.g. 30 nm. Interference filters achieve much narrower bandwidths by using optical interference between partially reflective surfaces separated by a dielectric film. A monochromator (see Section 16.2) is an optical system that will transmit a narrow bandwidth of wavelengths, $\Delta\lambda$, at a wavelength that can be adjusted by an external control.

Broad-band filters allow a specific *range of wavelengths* to pass, but will block all of the wavelengths outside of that range. Such filters can be used to remove the unwanted spectral wavelengths that are produced by diffraction-grating monochromators.

The *sinusoidal variation* in the spectral transmittance of a Michelson interferometer has a periodicity that can be controlled by the position of a movable mirror (Section 7.2.1). This is used in FTIR spectroscopic instruments to elicit information about the sample spectrum.

Calibration filters use materials with reproducible characteristics to provide *reference values* for either wavelength or transmittance.

DQ 13.13

What is the qualitative difference in the transmittance spectra of filters that are used for (i) wavelength calibration, and (ii) transmittance calibration?

Answer

Wavelength calibration requires a spectral structure that clearly defines specific wavelengths, i.e. sharp transmittance and absorbance peaks. Transmittance (or absorbance) calibration requires accurate transmittance values that do not change even if slight deviations in wavelength occur, i.e. they have a very slow spectral variation. Neutral-density filters provide a constant transmittance value over a wide wavelength range.

SAQ 13.4

An impulse signal has a very high intensity in a single pulse which has a very short duration. What would its Fourier transform look like?

SAQ 13.5

Estimate the minimum sampling frequency necessary to avoid *aliasing* when recording a Gaussian peak with a half-height width of 1 s. Assume that the tail of the Gaussian peak extends to **three** standard deviations from the centre of the peak (see also Figure 8.3 and equation (8.5)).

Note. Further questions that relate to the topics covered in this chapter can be found in SAQs 9.5, 10.3, 14.3 and 15.1.

Summary

This chapter introduced analogue and digital signals, and then developed the concept that analogue signals can be synthesized by a suitable summation of sine wave components. This then led to the reverse process of analysis of any wave into its frequency components, and the concept of the Fourier transform.

It then introduced the relationship between the use of the time and frequency domains to display the same signal information.

By using the randomness of noise-signal components, the discussion quantified the expected reduction in noise when passing through a low-pass filter, and then investigated the effect of analogue filters on 'noisy' signals.

Chapter 14

Unwanted Signals: Noise, Drift and Interference

Learning Objectives

- To differentiate between the characteristics of analytical signals, noise, drift and interference.
- To estimate the effect on signal-to-noise ratio of changing measurement bandwidth, detector temperature and signal strength.
- To appreciate the difference between the processes of signal filtering and signal integration.

14.1 Introduction

Can I tell the difference between a signal, a noise, drift, and interference?

In Section 13.1, we noted that the use of the word 'signal' normally implies the transfer of information that is wanted by the recipient. In this present section, we will use the term *signal* to apply to **all fluctuations** including both wanted *and* unwanted information. We will specifically identify the *wanted* signal by calling it the *analytical* signal.

All measurement processes produce *unwanted* signals in addition to the analytical signal. Different types of unwanted signals are often categorized as being *noise*, *drift* or *interference*. How can we tell the difference between them?

The process of classification depends on the following distinctions:

- an analytical signal carries **information** that is **wanted**;
- an interference signal carries **information** that is **not** wanted;

- noise or drift signals are **random** in character and carry **no** information;
- the **frequency range** of the **random** signal components determine whether they are classified as drift, long-term noise or short-term noise.

Interference is a fluctuation that is not random, because it has been induced by some other signal occurring in the outside environment. For example, a badly sited analytical balance may be subject to the vibrations induced by the operation of a nearby lift, or an aircraft navigation system may be affected by the signal from a mobile phone. The differentiation between an analytical signal and an interference signal is based on whether the information is wanted or not.

Noise and drift are both *random* fluctuations — they carry no information. The only differentiation between noise and drift lies in the following:

- our physiological perception of their frequency of fluctuation;
- the comparison with the frequency components of the analytical signal.

Rapid random fluctuations with a frequency higher than about 5 Hz are usually described as '(short-term) noise'. However, if the fluctuation appears to us to be *very slow*, with a period of more than about 100 s (i.e. less than 0.01 Hz), we would describe it as 'drift'.

Noise in the intermediate range (0.01–5.0 Hz), is often called 'long-term noise'. The fluctuation frequency is of the same order of magnitude as the frequency components in common analytical signals (e.g. a chromatogram). Hence, it is often difficult to distinguish long-term noise from a true analytical signal.

There are no exact frequency boundaries between short- and long-term noise and drift. The differentiation depends on the experimental system and the characteristics of the analytical signal that is produced.

DQ 14.1

The output from an instrument fluctuates randomly back and forth over a period of about 2 h. Estimate the frequency of the main component in this signal. Is this drift or noise?

Answer

The period, $T = 2$ h $= 2 \times 60 \times 60$ s.

Thus, the equivalent frequency, $f = 1/T = 1/(2 \times 60 \times 60) = 1.4 \times 10^{-4}$ Hz.

Such a slow variation would be called 'drift'.

The distinctions between the different types of signals are summarized in Table 14.1.

SAQ 14.1

A small bubble in an HPLC UV detector is causing some spurious signals on the chromatogram trace. Should we call this 'noise' or 'interference'?

Table 14.1 Distinctions between the different types of signals (wanted and unwanted)

Type of Signal	Carries information?	Wanted information?	Random?	Period, T Frequency f
Analytical	Yes	Yes	No	Any[a]
Noise (short-term)	No	No	Yes	T < analytical signal $T < \approx 0.2$ s, $f > 5$ Hz
Noise (long-term)	No	No	Yes	Similar to analytical signal
Drift	No	No	Yes	T > analytical signal $T > \approx 100$ s, $f < \approx 0.01$ Hz
Interference	Yes	No	No	Any

[a]Typical T ≈ 1–10 s.

14.2 The 'Noise' Spectrum

Not all 'noises' are the same.

All unwanted signals (noise, drift and interference) are often grouped within a loose generic concept of 'noise'. The 'noise' spectrum for a typical analytical instrument is illustrated in the *frequency-domain* graph presented in Figure 14.1. This is a plot of the amplitude of the 'noise' components as a function of their frequencies. Note that the frequency axis is drawn on a logarithmic scale.

The main features of this figure as follows:

(a) The overall noise envelope is the sum of various noise *factors*.

(b) White noise has components with *equal amplitude* at all frequencies, and is given the name 'white noise' in comparison with white light, which also has components at all wavelengths. White noise is normally the most significant noise factor at frequencies above about 100 Hz. In analytical

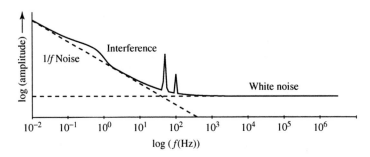

Figure 14.1 'Noise' spectrum of a typical analytical instrument showing the various noise components (for the frequency domain).

instrumentation, there are two important sources of white noise — thermal noise (see Section 14.4.1) and shot noise (see Section 14.4.2).

(c) At low frequencies (below about 100 Hz), the most important noise factor is '$1/f$ Noise' (see Section 14.3). The term for this 'noise' derives from its frequency response, where the amplitude is inversely proportional to frequency.

(d) The effects of some typical sources of interference are shown on the figure. A very common source of interference is electromagnetic pick-up from the mains power supplies at 50 and 100 Hz (60 and 120 Hz in the USA).

The total amount of noise that is present in a particular instrument system will depend on the range of frequencies to which the system responds, with the response range of a system being called its *bandwidth*, Δf (see Section 13.6). For example, a system that can respond to frequencies from 30 to 230 Hz has a bandwidth of 200 Hz. Such a system could expect to encounter $1/f$ noise at lower frequencies ($30-(ca.)100$ Hz), white noise at all frequencies within the range, and mains supply interference at 50 and 100 Hz.

If the bandwidth of the system can be reduced, then the total amount of noise present can also be reduced. This is a very important way of reducing random noise (Section 13.6.1).

DQ 14.2

A new instrument system will use a signal of frequency, f, to transfer information. Which would be the best frequency to use to maximize the signal-to-noise ratio:

(a) 10 Hz;
(b) 100 Hz;
(c) 1 kHz;
(d) 10 kHz?

Answer

At a low frequency, we can expect that the effect of $1/f$ noise will increase — hence the low frequency (a) of 10 Hz would not be a good choice. At 100 Hz, we might also expect interference from an harmonic of the (UK) mains frequency of 50 Hz — a frequency of 1 kHz would therefore be a better choice.

Would 10 kHz be even better than 1 kHz? In fact, above about 100 Hz the main source of noise is white noise, which has the same magnitude at all frequencies — so (c) and (d) are probably equally valid answers.

14.3 1/f Noise

When is a 'noise' not noise?

There are many factors that contribute to $1/f$ ('one over f') noise, both at the level of the electronic device and at the level of the instrument. One source, as an example, may be due to the effect of minute amounts of moisture evaporating from high-resistance electronic components, thus resulting in a gradual change in resistance.

However, the main *characteristic* common to all $1/f$ factors is the decrease in amplitude as a function of the frequency at which it is measured. This noise has also been called 'Flicker Noise'.

At low frequencies, $1/f$ noise is equivalent to the familiar concept of *drift*. As expected, the *effect* of drift increases with the time available for the drift to occur. For example, the drift in the 'zero absorbance' (0A) setting of a spectrophotometer may be quoted as 0.001A per hour, thus giving a possible error that increases with time. As the effective frequency is inversely proportional to the time, T, of the drift, we can therefore write the following:

$$\text{Magnitude of Drift} \propto 1/f \qquad (14.1)$$

DQ 14.3

The drift in a UV–visible spectrophotometer is quoted as 0.01A/h. Estimate the possible percentage error in the measurement of the concentration of an analyte if the true absorbance of this analyte is 0.05A and the times between measurement of the test and reference samples are as follows:

  (i) 30 min;
 (ii) 2 min;
(iii) 0.02 s.

Answer

The possible error in absorbance, ΔA, is given by 0.01 × Δt, where 't' is the time delay in hours — this gives (i) 0.005, (ii) 3.33 × 10^{-4} and (iii) 5.56 × 10^{-8}. The percentage error in concentration will be the same (see equation (5.11)) as the percentage error in A(= 100 × (ΔA)/A), thus giving (i) 10%, (ii) 0.67%, and (iii) 0.00011%.

The lessons to be learnt from the above discussion question are that the effects of drift can be (i) *most significant* when the *analytical signal is itself small*, and (ii) *dramatically reduced* by reducing *the time between calibration and measurement* — see Section 14.5.1.

14.4 White Noise

Why 'white'?

The key features of 'white' noise are as follows:

- it has components with *equal amplitude* at all frequencies (as white light);
- these frequency components are in *random* phase.

The total noise observed in an instrument will be the 'sum' of all of the separate noise components. Hence, if an instrument has a larger bandwidth, Δf, it will be able to respond to a wider range of noise frequency components — the *total* noise signal will therefore be higher.

However, each frequency 'component' in the spectrum will have a random phase compared to all of the other frequency components. We have seen earlier in Section 13.6.1 that the total white-noise signal amplitude in an instrument is given by the following expression:

$$S_{N,\Delta f} = \sqrt{(\Delta f)} \times S_{N,1} \tag{14.2}$$

where Δf is the bandwidth of the instrument, and $S_{N,1}$ is the white-noise signal amplitude for unit bandwidth.

14.4.1 Thermal (Johnson) Noise

Thermal noise occurs in all electronic components, including detectors. Except at the absolute zero of temperature, the electrons (and holes) will have thermal kinetic energies which create random motion of the electrical charge. Statistical fluctuations in this charge distribution will thus lead to a random potential difference appearing across the terminals of the component. The rms voltage, $V_{N,rms}$, generated by thermal noise is given by the following:

$$V_{N,rms} = \sqrt{(4kTR\Delta f)} \tag{14.3}$$

where Δf is the bandwidth of the system that is recording this voltage, R is the resistance of the conductor, T is the temperature (in K), and k is the Boltzmann constant.

14.4.2 Shot Noise

Electric current is measured as the charge passing a given point at a given instant (1 amp (A) = 1 coulomb (C) passing per second (s)). However, the charge is carried by *individual* electrons (each with a discrete charge of -1.6×10^{-19} C), and is not uniformly distributed in the same way as, e.g. water flowing down a pipe. The passage of individual electrons gives rise to fluctuations in the current flow.

The rms current noise, $I_{N,rms}$, is given by the following:

$$I_{N,rms} = \sqrt{(2I_0 e \Delta f)} \tag{14.4}$$

where Δf is the bandwidth of the system that is recording this current, e is the charge on the electron, and I_0 is the steady current through the device.

SAQ 14.2

A photomultiplier tube, measuring the light intensity in a spectrophotometer, produces an electric current which is proportional to this intensity. The current signal is converted into a voltage signal by passing the current through a resistor. As the current is very small, the resistor must have a high value in order to develop a reasonable voltage. Assume the following:

value of the resistor, $R = 500$ MΩ;
signal current, $I = 1.0$ pA $= 1.0 \times 10^{-12}$ A;
temperature of the resistor, $T = 25°C = 298$ K;
detection bandwidth $= 100$ Hz.

(i) Estimate the magnitude of the voltage signal, and (ii) identify the possible sources and magnitudes of uncertainties in this signal.

14.4.3 Comparison of Thermal Noise and Shot Noise

Both thermal noise and shot noise are examples of *white noise*, and hence both show the characteristic dependency on the square root of the bandwidth, $\sqrt{(\Delta f)}$.

The amplitude of the thermal noise, V_N, depends on the temperature, T, of the electronic component, but it does *not* depend on the magnitude of the analytical signal. Conversely, the amplitude of the shot noise, I_N, does not depend on the temperature of the component, although it *does* depend on the magnitude of the analytical signal — the current, I_0.

Note that thermal noise could be reduced by lowering the temperature of the component. This fact is often used in the detection of IR radiation, when the detector may be cooled in order to improve the signal-to-noise ratio.

DQ 14.4

A detection system with a bandwidth of 250 Hz has an S/N ratio of 12 when the detector is at room temperature (25°C). Calculate the new S/N ratio if the detector is cooled with liquid nitrogen ($-196°C$) and the bandwidth is reduced to 10 Hz, for (i) shot noise, and (ii) thermal noise.

Answer

(i) *Shot noise will be unaffected by the drop in temperature, but will be reduced as the square root of the reduction in bandwidth, i.e.* $\sqrt{(250/10)}$, *thus giving a new S/N ratio of* $12 \times 5 = 60$.

(ii) *Thermal noise will display the same improvement due to bandwidth, but will also increase due to the drop in temperature, i.e. $S/N = 12 \times 5 \times \sqrt{(298/77)} = 118$.*

DQ 14.5

A detector has an S/N ratio of 12 for a given signal amplitude. Calculate the new ratio if the signal amplitude is increased by a factor of 4, for (i) shot noise, and (ii) thermal noise.

Answer

(i) *The signal current, I_0, will be proportional to the signal amplitude. Shot noise will increase as the square root of the current $(= \sqrt{4})$, so that the increase in the signal-to-noise ratio will be $4/\sqrt{4} = \sqrt{4}$. The final S/N ratio will be $12 \times \sqrt{4} = 24$.*

(ii) *Thermal noise will be unaffected by the signal level, so the S/N ratio will increase in direct proportion to the signal strength. The final S/N ratio will be $12 \times 4 = 48$.*

14.5 Reduction of Noise, Drift and Interference

How can an instrument differentiate between an unwanted signal and a wanted analytical signal?

We have seen in Section 12.4 that all signals are the sum of a set of frequency components, with each frequency component being a pure sine wave that can be described entirely by the characteristics of amplitude, frequency and phase. However, there is no characteristic that can identify a frequency component as specifically 'belonging to' the analytical signal. An instrument system will treat a 10 Hz *noise* component in exactly the same way as a 10 Hz *analytical* component.

If an instrument system is to be able to distinguish between signals, it is necessary to make use of additional information that will enable such a distinction to be made. This necessary information is given above in Table 14.1.

14.5.1 Reducing the Effect of Drift

Drift is a slow change in a constant (DC) signal. It is not possible to eliminate drift entirely, although developments in modern electronics have considerably reduced drift in the electronic systems of analytical instruments.

DQ 14.3 illustrated the effect of drift in the zero-absorbance setting (0A), i.e. the baseline drift, of a spectrophotometer. The *effect* of such a drift can be minimized if the time between setting 0A and making the measurement can be made as short as possible. The *double-beam* spectrophotometer (Section 5.3) reduces this comparison period very considerably by using a rotating mirror to 'chop' rapidly between the reference sample and the test sample.

The process of 'chopping' changes the frequency at which the signal is measured in the system, and effectively 'moves' the analytical signal to a 'quieter' part of the frequency spectrum where the noise is less.

DQ 14.6

Noise in a spectrophotometer could be reduced by a factor of 2 by increasing the chopping frequency from 400 to 800 Hz — true or false ?

Answer

According to equation (14.1), we might expect the effect of drift to be reduced by a half. However, Figure 14.1 shows that above (very approximately) 100 Hz, the effect of white noise becomes more significant than drift. As white noise has the same amplitude at all frequencies, there will be no additional advantage gained by 'chopping' at higher frequencies.

14.5.2 Reduction of Interference

Interference can be a very real problem because there is no generic way to distinguish it from our analytical signal — both the analytical signal and interference signal carry *information*.

Some examples of this include the mains frequency, 50 Hz (60 Hz in the USA) and its harmonic 100 Hz (120 Hz in the USA), the radiation from mobile phones, and switching transients on the power grid. Where possible, such interference should be prevented through the use of electronic filters and adequate electromagnetic screening. Other interference can occur through mechanical vibrations, and care must be taken to site equipment, e.g. balances, correctly.

DQ 14.7

Which of the following GC detectors are most likely to be affected by electromagnetic interference: (a) a thermal-conductivity detector; (b) a flame-ionization detector; (c) a flame-photometric detector; (d) an electron-capture detector?

Answer

The detectors most sensitive to electromagnetic interference will be those that measure very small electric currents. Such detectors may be susceptible to the small currents induced by external fields. If necessary, refer back to Section 10.3 to identify them.

If it is not possible to eliminate the interference, then it is necessary to attempt to reduce its effect. The reduction of mains interference can be achieved by using the knowledge of its exact frequency of 50 Hz (60 Hz (USA)), and employing a notch filter (see Figure 13.8(d)) in the signal path to absorb that particular frequency component. Note that in this case it is important to ensure that the instrument is designed for the particular frequency of the mains supply being used, i.e. an instrument designed for 50 Hz would not provide maximum interference rejection on a 60 Hz supply.

14.5.3 Analogue Filters

The use of a low-pass analogue filter for reducing noise, by removing the frequency components due to unwanted noise, was introduced previously in Section 13.6. However, this type of filter can not distinguish between noise components and signal components, so a filter bandwidth, Δf, must be chosen that allows all (or almost all) of the signal components to pass, but then absorbs the higher-frequency noise components.

DQ 14.8

Which of the following happens to a chromatographic peak if it is passed through a low-pass filter with a time constant that is of a similar magnitude to the width of the peak:

(a) the peak width is increased;
(b) the peak height is reduced;
(c) the peak position is delayed in time?

Answer

*In this case, many of the frequency components of the signal will fall outside of the bandwidth of the filter. The effect will be that the signal peak will be severely distorted, and all three options will occur. Not only does the low-pass filter smooth out the signal itself, but its sluggish response also means that the signal peak is delayed in **time**.*

It is clear from the above discussion question that it is necessary to establish a compromise between smoothing high-frequency noise from the chromatogram and distorting the peaks. In practice, the width of the peak should be at least 10 times the time constant of the filter in order to avoid distortion.

SAQ 14.3

A GC peak of width 0.1 min is observed with a S/N ratio of 10, when the bandwidth of the low-pass output filter is 100 Hz. What are the new values of S/N when the bandwidth is decreased to (a) 10 Hz, (b) 1 Hz and (c) 0.1 Hz?

14.5.4 Digital Filter

The use of convolution in digital filters is introduced later in Section 15.4. A digital filter, using convolutes, can reduce noise, but without significantly distorting the peak or shifting it in time.

DQ 14.9

Is it possible to build an analogue filter that can perform the same function as the Savitsky and Golay convolutes (see Section 15.4), i.e. smooth out the noise but retain the shape and position of the spectral line?

Answer

No — designing 'digital filters' by using computer software permits methods of signal processing that are not possible in the 'real-time' analogue world.

14.5.5 Signal Averaging and Integration

Signal averaging is a process whereby several data points are averaged together in order to provide a single data point. This technique is used to improve the S/N ratio in a variety of ways, with examples including the following:

- wavelength bunching in diode-array detectors;
- repeated scans in FTIR spectrophotometry;
- signal integration in (flame) atomic absorption spectroscopy.

The analytical signal increases in proportion to the *time* of the integration, T, or the *number* of repeated measurements, n, as given by the following:

$$S_{A,n} = nS_{A,1} \tag{14.5}$$

The *random*-noise signal increases in proportion to the square root of the time of the integration, $\sqrt{T}$, or the square root (see equation (13.7)) of the number of repeated measurements, $\sqrt{n}$, as follows:

$$S_{N,n} = \sqrt{n}S_{N,1} \tag{14.6}$$

Therefore, the signal-to-noise ratio increases according to the following relationship:

$$(S/N)_n = \frac{S_A}{S_{N,n}} = \frac{n \times S_A}{\sqrt{n} \times S_{N,1}} = \sqrt{n} \times \frac{S_A}{S_{N,1}} = \sqrt{n}(S/N)_1 \tag{14.7}$$

and hence S/N is proportional to $\sqrt{n}$ or $\sqrt{T}$.

An *integrator* averages a 'constant' analytical signal over a period of time, T, and then gives the result at the end of that period. The operator must therefore wait for the end of the next integration period, T, before the value can be updated.

The use of computer memory in spectrophotometers now permits the averaging of repeated scans. Each wavelength data point is saved separately in the memory, and is then averaged with equivalent data points from subsequent repeated spectral scans.

DQ 14.10

In a particular (flame) atomic absorption spectrophotometer, the output signal can be either (a) smoothed by using a low-pass filter with a time constant, τ, or (b) integrated over a period T.

Which method should be used when carrying out the following:

 (i) tuning the instrument for maximum signal response;
(ii) taking the analytical reading?

Answer

*When tuning the instrument, the operator needs to be able to see changes in response **immediately**, e.g. when adjusting the height of the flame. The integrator would not be suitable for this, as the operator has to wait for the end of the integration period before any changes in response become apparent. The low-pass filter, however, will show some response to change immediately, and indicate whether the adjustment is going in the right direction. When the actual signal is being measured under stable conditions, the integration mode is appropriate as it gives a value that is the average over the desired period.*

SAQ 14.4

In a particular HPLC diode-array detector, the intensity signal is read from each of the photodiodes at a rate of 100 times per second. A number, (n), of sequential values from **each** photodiode are then 'averaged' together to produce a single data value. The operator can set the value for 'n'. Which of the following statements are true?

(a) Increasing the value of 'n' will improve the S/N ratio.

(b) The maximum useful value for 'n' depends on the chromatograph peak width.

(c) Increasing the value of 'n' will slow down the recording of the chromatogram.

(d) The maximum useful value for 'n' depends on the number of diodes in the array.

Note. Further questions that relate to the topics covered in this chapter can be found in SAQs 2.1, 6.3, 7.2, 18.1 and 18.3, and DQs 6.7, 7.5, 10.6, 13.11, 13.12 and 15.3.

Summary

This chapter has discussed the main types of unwanted signals, namely noise, drift and interference. It demonstrated how an understanding of the origins of each of these can be used in order to reduce the unwanted noise and its effects.

The processes of filtering are dealt with elsewhere in this text, i.e. analogue and digital filters in Chapter 13 and 15, respectively, but this present chapter was used to examine the general concept of filtering in comparison to the process of signal-averaging as mechanisms for improving signal-to-noise ratios.

Chapter 15

Convolution

Learning Objectives

- To predict the result of the convolution of simple functions.
- To explain the origin of triangular and sinc instrument lineshapes for UV–visible and FTIR spectrophotometers, respectively.
- To describe the operation of a digital filter by using the concept of convolution.
- To demonstrate why a wide spectral bandwidth can lead to a non-linear photometric response for a UV–visible spectrophotometer.

15.1 Introduction

Convolution is a mathematical procedure that calculates an interaction between one function and another. This process can be used to help the understanding of certain practical situations (e.g. instrument lineshapes), and we will use one of these practical examples as a way of *introducing* the concept of convolution.

Convolution also plays a very important role in the computing procedures used in the software signal processing of analytical signals. This is developed later in Section 15.4 on *digital filtering*.

15.2 Case Study: the Cell Volume of an HPLC Detector

Does a larger cell volume indicate greater sensitivity or greater peak broadening?

We will introduce the concept of convolution by considering the effect that the *cell volume* of a UV–visible detector might have on the observed width of an HPLC peak.

Figure 15.1 shows the profile of an HPLC absorbance peak, given as a function of time. The peak used in this example has a retention time of 3.90 min, a peak width (at half-height) of 0.2 min, and a flow rate of 1.5 ml/min. Note that we express subdivisions of a minute here by using *decimal fractions* and *not* seconds.

The analyte signal is being carried through the detector by the mobile phase at a predetermined flow rate, V_F, (expressed in units of ml/min). Hence, it is also possible to calibrate the horizontal scale in Figure 15.1 both in terms of the *time t*, and the *volume V* of the fluid passing through the detector. For example, 3.90 min is equivalent to $3.90 \times 1.5 = 5.85$ ml, and 0.20 min is equivalent to $0.20 \times 1.5 = 0.30$ ml.

The UV–visible detector records the absorbance of the eluent as it passes through a detector cell of volume V_A (Figure 15.2). The peak in Figure 15.1 has been drawn by assuming that this detector volume is *so small* that its effect on the shape of the peak can be neglected. Figure 15.1 can therefore be considered to give the 'true' absorbance profile of the peak.

The absorbance values of the peak are sampled (Section 13.2.2) at intervals of 0.02 min (0.03 ml), as indicated by the 'points' on the plot. The absorbance values occurring between 3.80 and 4.08 min are also reproduced below in Table 15.1 (column 3).

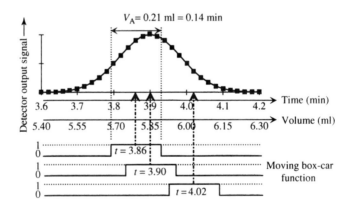

Figure 15.1 Effect of the detector cell volume on the observed width of an HPLC absorbance peak.

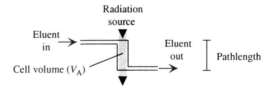

Figure 15.2 Schematic representation of the UV–visible detector cell.

Table 15.1 The process of convolution

Retention time (min)	Equiv. volume (ml)	'True' absorbance	Time (min)					
			3.86		3.90		4.02	
			Cell value[a]	Absorbance	Cell value[a]	Absorbance	Cell value[a]	Absorbance
1	2	3	4	5	6	7	8	9
3.80	5.70	0.500	1	0.500	0	0.000	0	0.000
3.82	5.73	0.642	1	0.642	0	0.000	0	0.000
3.84	5.76	0.779	1	0.779	1	0.779	0	0.000
3.86	5.79	0.895	1	0.895	1	0.895	0	0.000
3.88	5.82	0.973	1	0.973	1	0.973	0	0.000
3.90	5.85	1.000	1	1.000	1	1.000	0	0.000
3.92	5.88	0.973	1	0.973	1	0.973	0	0.000
3.94	5.91	0.895	0	0.000	1	0.895	0	0.000
3.96	5.94	0.779	0	0.000	1	0.779	1	0.779
3.98	5.97	0.642	0	0.000	0	0.000	1	0.642
4.00	6.00	0.500	0	0.000	0	0.000	1	0.500
4.02	6.03	0.369	0	0.000	0	0.000	1	0.369
4.04	6.06	0.257	0	0.000	0	0.000	1	0.257
4.06	6.09	0.170	0	0.000	0	0.000	1	0.170
4.08	6.12	0.106	0	0.000	0	0.000	1	0.106
Sum of values				5.761		6.294		2.822
Normalized sum (= apparent absorbance)				0.823		0.899		0.403
'True' absorbance				0.895		1.000		0.369

[a] '1', fluid inside the cell; '0', other data points.

In practice, the volume, V_A, of the detector cell is not zero. The following analysis demonstrates the effect that an increase in this detector volume would have on the observed absorbance profile of the peak.

We will investigate, as an illustrative example, the effect on the peak if the detector cell were to have a 'large' volume equal to 0.21 ml. The fluid will then take 0.14 (= 0.21/1.5) min to travel through this cell.

Consider the situation at time $t = 3.86$ min — the volume of fluid *in the cell* at this time corresponds to the peak section of width V_A, as shown in Figure 15.1. This means that the analyte component with a retention time of 3.93 min has already entered the cell before the analyte with a retention time of 3.79 min has left the cell. The observed absorbance recorded by the detector will then be the *sum* of all of the absorbances of the components within the range V_A.

In order to compare the situation for different cell sizes, we divide the total absorbance value by a normalizing factor. This then gives the *average* of the absorbances of the components within the range V_A. Table 15.1 shows, for example, how this can be done at three different times, i.e. 3.86, 3.90 and 4.02 min.

As the fluid passes through the cell, the volume, V_A, will cover a different section of the peak. The plot of the *average* absorbances for different times will still give the *observed* peak.

The columns given in Table 15.1 show the following information:

Column 1   the *retention times*;
Column 2   the *equivalent volumes*;
Column 3   the *'true' absorbance* value;
Columns 4 and 5   data for *calculating* the *observed absorbance* at $t =$ 3.86 min;
Columns 6 and 7   data for *calculating* the *observed absorbance* at $t =$ 3.90 min;
Columns 8 and 9   data for *calculating* the *observed absorbance* at $t =$ 4.02 min;
Columns 4, 6 and 8   *'cell values'*, where '1' indicates which data points correspond to fluid inside the cell, and '0' represents all other data points.

The cell volume (0.21 ml) covers *seven* data points in the peak profile. Hence, the absorbances of seven sample points will be averaged together in the cell at any given time.

The *process* of selecting which seven points are averaged, and then carrying out the process, can be automated in a simple mathematical way. As can be seen in Table 15.1, at $t = 3.86$ min, column 4 shows a '1' at the retention times for which the analyte component will be in the cell. All of the other retention times will show a '0' in this column. This form of data presentation is often called a 'box-car' function, and is also shown in Figure 15.1. The flow of fluid through the cell is simulated by moving the 'box-car' along the time axis in order to 'sample' the different sets of seven data points.

The data presented in column 5 are generated by multiplying each value in column 3 by the equivalent value in column 4 (either 0 or 1), a process known as *multiplying* on a 'point-by-point' basis. It is obvious that only the seven relevant retention times will give a combined value that is not zero. The next step is to add all of the values in column 5 to produce a single value, i.e. 5.761. This is the mathematical process of *integration*.

Finally, the sum must be divided by *seven* in order to obtain the necessary *average* value, i.e. 0.823. This is the mathematical process of *normalization*.

The *observed* average absorbance for $t = 3.86$ min is therefore calculated to be 0.823. The 'true' value of the absorbance that would have been obtained for a very small cell volume at $t = 3.86$ min is 0.895 (see column 3).

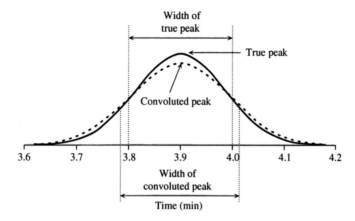

Figure 15.3 Comparison of the 'true' and convoluted peaks.

The whole process of *multiplication, integration,* and *normalization* can then be repeated for the two other values of the retention time.

At $t = 3.90$ min, the box-car function in column 6 has moved to the new position shown in Figure 15.1, where it 'accepts' data from the seven points which straddle the peak. The observed absorbance is 0.899, which is 10% *less* than the 'true' value of 1.00.

At $t = 4.02$ min, the box-car function for column 8 has moved to the next position shown in Figure 15.1. In this case, it can be seen that the observed absorbance, i.e. 0.403, is now actually *higher* than the 'true' value of 0.369.

Figure 15.3 compares the 'true' absorbances with those obtained by repeating the above process for every data point — the curve obtained for the latter is known as the 'convoluted' peak. It can be seen that the effect of the cell *volume* is to spread out the peak, thus reducing the relative absorbances in the central region, but increasing the absorbances in the wings.

For ease of comparison, we have used a normalizing factor such that the *areas* of the true and broadened peaks have remained the same. In practice, the area of the absorbance peak will be proportional (assuming Beer's Law) to the optical *pathlength* in the detector.

DQ 15.1

The area of the HPLC peak discussed above is measured to record the total amount of analyte present, with a flow cell with a longer pathlength and a larger volume being used to increase the responsivity. Which of the following statements are true, possibly true, or false?

(i) The responsivity of the detector is a function of the cell volume.

(ii) The broadening of the peaks is a function of the cell volume.

Analytical Instrumentation

(iii) A slight broadening of the peaks does not affect the accuracy of the results.

(iv) The optimum design of a cell would give a minimum volume with a maximum pathlength.

Answer

 (i) *Possibly true — responsivity is a direct function of the **pathlength**, but a cell with a larger volume will normally also have a longer pathlength.*

 (ii) *True.*

(iii) *Possibly true — the **area** of the peak is independent of the broadening, but there may be a significant second-order effect in that there may be overlap with the tails of other peaks that are also broadened.*

(iv) *True.*

15.3 Steps in the Process of Convolution

A recipe for mixing the ingredients of a convolution would be useful.

By referring to our introductory example in the previous section, we can identify the general steps involved in the convolution process (see Table 15.1). These are as follows:

(1) The two functions, F_1 (the peak in Figure 15.1) and the convoluting function, or *convolute*, F_2 (the box-car in Figure 15.1), are multiplied together point-by-point to obtain a set of *intermediate* data values[†].

(2) These intermediate values are integrated (summed) and then normalized to give a *single* output value.

(3) The convolute F_2 is *displaced* (moved) with respect to the other function, and the whole multiplication process is then repeated in order to obtain a new *output value*.

(4) The final function, F_O, is the *set of the output values* generated as a function of the displacement between the two initial functions.

This is written mathematically as follows:

$$F_O = F_1 \otimes F_2$$

[†] Note that mathematically the convolute function, F_2, must be reversed before carrying out the convolution process. We do not notice the effect of this in our discussion here, because all of the convolutes used are symmetrical and reversal will not change the result.

15.3.1 *Use of the Fourier Transform in Convolution*

It is also possible to *calculate* the result of the convolution of two functions, $F_1 \otimes F_2$, by a point-by-point multiplication of the Fourier transforms of the individual functions. This process can be summarized as follows:

(1) The Fourier transforms, f_1 and f_2, of the two functions F_1 and F_2, respectively, are calculated.

(2) The functions f_1 and f_2 are multiplied together, point-by-point, to obtain an intermediate function, $f_0 (= f_1 \times f_2)$.

(3) The inverse Fourier transform of f_0 then gives the final function, i.e. F_0.

A good example of the use of this relationship occurs in the development of the sinc lineshape (Section 7.4.2) for the FTIR spectrophotometer.

SAQ 15.1

Explain the **relationship** between the following **two statements**:

(i) The spectral bandwidth, due to a monochromator with entrance and exit slits of equal widths, has a *triangular* shape.

(ii) The product of a sinc (x) function multiplied by itself is a $sinc^2$ (x) function.

15.4 Smoothing Convolutes — Digital Filters

In our introductory example using the detector cell, we performed a convolution operation on the digitized chromatogram in order to demonstrate a *physical* effect. However, convolution is also widely used in many areas of digital signal- and image-processing. As a mathematical technique in *software applications*, it is used to accentuate, or dampen, different aspects of the information contained in a signal — digital filtering.

In each of the following examples, we will develop a new 'convolute'. This is a set of numbers that provides the 'moving' function, F_2, with which we multiply, point-by-point, our original data function, F_1. Each convolute can be normalized, i.e. each value can be multiplied by a single factor, such that the *sum* of the convolute is unity.

The simplest application is the *box-car smoothing filter*, which is used to improve the signal-to-noise ratio. In this process, a number, n, of adjacent data values are averaged together in order to provide a single output value. This process is called 'box-car averaging' in the sense that it represents a 'rectangular railway truck' moving across the spectrum, collecting and averaging data. The box-car convolute used in Table 15.1 appears in columns 4, 6 and 8, and has the values $\{1, 1, 1, 1, 1, 1, 1\}$ The normalizing factor for this convolute is $1/7 = 0.143$.

We have seen earlier in Section 14.5.5 that the effect of averaging 'n' data points is to improve the *signal-to-noise ratio* in the aggregate value by a factor of $\sqrt{n}$, as given by the following:

$$(S/N)_n = \frac{S_A}{S_{N,n}} = \frac{n \times S_A}{\sqrt{n} \times S_{N,1}} = \sqrt{n} \times \frac{S_A}{S_{N,1}} = \sqrt{n}(S/N)_1 \qquad (15.2)$$

Although it provides a very simple method of digital smoothing, the box-car method has severe limitations when the width of the box-car is significant when compared to the maximum linewidth in the spectrum. We have seen above that a cell of equivalent width 0.14 min reduces the apparent peak height of a chromatogram line of width 0.2 min to 90% of its true peak height. It also significantly broadens the line.

The box-car smoothing convolute has the following 'properties':

- it does not shift the apparent position of the line;
- it does not affect the overall area of the line;
- it *does*, however, affect the linewidth of the line — significant broadening can occur.

Savitzky and Golay [1] developed sets of smoothing convolutes that produce less distortion. The mathematics of these convolutes is based on procedures for fitting a polynomial curve to the data points.

Figure 15.4 shows the effect of digital smoothing on a spectral line of width 11 nm with sampling at 1 nm intervals, when it is smoothed by using the following:

(a) a box-car function of width 11 nm;

(b) the 11-point (Savitzky–Golay) convolute:
   $\{-36, 9, 44, 69, 84, 89, 84, 69, 44, 9, -36\}$.

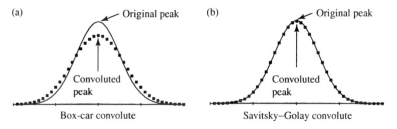

Figure 15.4 Digital smoothing of a spectral line (Gaussian peak) by using (a) a box-car function, and (b) a Savitzky–Golay convolute.

The box-car convolute causes line broadening and a drop in peak height. However, the use of the Savitzky–Golay convolute averages the noise across 11 points, but still does not significantly distort the line.

Note that, with a 'moving-convolute' digital filter, the density of the data points in the smoothed spectrum is the same as the density in the original unsmoothed spectrum. This can be contrasted with the bunching of signals from separate diodes in a diode-array detector (Section 9.3.2), where a single averaged value replaces several data values.

It is also possible to use the process of convolution to carry out a *differentiation* of the signal, and appropriate convolutes have been produced [1].

DQ 15.2

Apply a triangular convolute {1, 2, 1} with a normalizing factor of 0.25 to the following set of data points:

0.2   0.5   1   2   4   5   4   2   1   0.5   0.2

Which of the following is the correct result?

(i) 2.2	4.5	9	15	18	15	9	4.5	2.2		
(ii) 0.05	0.125	0.25	0.5	1	1.25	1	0.5	0.25	0.125	0.05
(iii) 0.55	1.125	2.25	3.75	4.5	3.75	2.25	1.125	0.55		
(iv) 3.5	13	3.5								

Answer

Answer (i) has applied the convolute but has not included the normalizing factor, while answer (ii) has been obtained by simply multiplying the original data by the normalizing factor.

Answer (iv) has taken groups of three data points and added them together to get three points. This is the process of data bunching (Section 9.3.2). It has not used a convolute that moves along one data point at a time.

Answer (iii) is the correct answer.

There is one note of caution to be made concerning the use of smoothed data. Once the data have been filtered by using a convolution process, they are no longer *raw* data. For the purposes of traceability, the original data must be retained separately and should always be used as the starting point for subsequent statistical analysis.

DQ 15.3

(i) Does a low-pass analogue filter cause a time shift in the position of the peak?

(ii) Does a digital filter cause a time shift in the position of the peak?

Answer

A low-pass analogue filter has the effect of broadening the line and delaying the time of the peak maximum (see DQ 14.8). However, the amount of delay is not constant, and depends on the signal linewidth.

*The digital convolute can act in pseudo 'real-time', i.e. it can carry out the calculations as the data arrive, moving, point-by-point, from one set of data points to the next. There will be a small time delay, because the convolute of 'n' data points can not be calculated until n/2 data points following the mid-point of the set have been received. However, because the data are processed by using computer software, an exact correction can be made on the time scale, so that there is **no shift** in the position of the peak as it is finally presented.*

15.5 Instrument Lineshape

Do all instruments have the same lineshape?

The 'instrument lineshape' (ILS) is the spectral output that would be produced by a spectrophotometer in response to a theoretical monochromatic (or very narrow) spectral line.

We will see in Section 16.2.3 that a diffraction-grating monochromator normally produces a triangular lineshape which is dependent on the widths of the entrance and exit slits. In fact, it is possible to understand the origin of the triangular lineshape as the *convolution* of the following two functions.

(a) F_1 — the rectangular function that is the image of the entrance slit, produced by the diffraction grating;

(b) F_2 — the rectangular function that is the opening of the exit slit.

The convolution of two rectangular functions of the same widths is a triangular function with a width (at half-height) which is equal to the width of the rectangular functions (see Figure 16.5).

The basic ILS of an FTIR spectrophotometric instrument is a sinc function, which is the Fourier transform of the rectangular function imposed by limited mirror travel on the interferogram (Section 7.4.3).

15.5.1 Effect of Instrument Lineshape

The instrument lineshape (ILS) acts as a convolute, F_2, applied to the true spectrum, F_1. At each wavelength setting in the spectrum, the observed signal will be the *summation* of components from across the spectral bandwidth of the instrument, with each component being multiplied by the magnitude of the ILS function at that particular wavelength. This is equivalent to the first two steps in the convolution process (see Section 15.3 above). Scanning F_2 through the spectrum F_1 carries out the next stage in the convolution process to produce the final *observed* spectrum, F_O.

DQ 15.4

In calculating the effect of a triangular ILS in a UV–visible spectrophotometer, does it matter whether we convolute the ILS with the **transmittance** spectrum or with the **absorbance** spectrum?

Answer

*The instrument produces a **transmittance** spectrum directly, from which the absorbance spectrum is only **calculated** (Section 5.1). Thus, the ILS acts as a convolute for the transmittance spectrum.*

The above discussion question shows that in order to calculate the effect of a triangular transmittance ILS on an *absorbance* spectrum (see Figure 15.5), it is necessary to carry out the following:

- convert the absorbance function back into a *transmittance* function by using $T = A^{-10}$;
- *convolute* the transmittance spectrum with the triangular lineshape;
- convert the result back to an absorbance spectrum, by using $A = -\log T$.

Figure 15.5(a) shows the *reduction in peak absorbance* $(A_P - A_0)$ due to the convolution process of a triangular bandpass with the *transmittance* spectrum. The reduction is expressed as a percentage of the *observed* value $= 100 \times (A_P - A_0)/A_0$ and is presented as a function of both of the following:

- the ratio, Δ_I/Δ_S, of the spectral bandwidth Δ_I to the analyte linewidth, Δ_S;
- the peak absorbance, A_P, of the sample.

Some important points to note about Figure 15.5(a) are as follows:

- the percentage reduction increases for wider spectral bandwidths;
- the percentage reduction increases for peaks with greater absorbance.

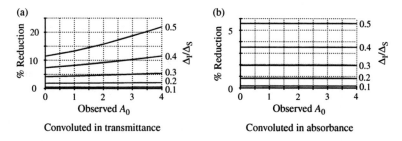

Figure 15.5 Effect of convolution with a triangular bandpass, carried out on (a) the transmittance spectrum, and (b) the absorption spectrum.

Figure 15.5(b) shows the effect of the convolution of a rectangular bandpass with the *absorbance* spectrum. Again, the reduction is expressed as a percentage of the *observed* value $(= 100 \times (A_P - A_0)/A_0)$ and is presented as a function of both of the following:

- the ratio, Δ_I/Δ_S, of the spectral bandwidth Δ_I to the analyte linewidth, Δ_S;
- the peak absorbance, A_P, of the sample.

DQ 15.5

The peak absorbances of the same spectral line are measured for both a **test** sample and a **standard** sample. The spectral bandwidth for the instrument is 5 nm, while the analyte linewidth is only 10 nm. If the observed absorbances for the standard sample and test sample are $A_R = 0.5$ and $A_T = 3.0$, respectively, calculate the error in estimating the concentration of the test sample.

Answer

The concentration of the test sample is given by $c_T = (A_T/A_R) \times c_R$, giving an apparent value of $6c_R$. Using the '0.5' line in Figure 15.5(a) for the ratio of Δ_I/Δ_S, we can see that A_T (at 3.0A) is 19% too small and A_R (at 0.5A) is 12% too small. This would give a true value for $(A_T/A_R) = 6 \times 1.19/1.12 = 6.38$. Thus, the true value for the concentration will be over 5% more than the apparent value.

DQ 15.6

In DQ 15.5, we did not use the 'combination of uncertainties' relationship derived in Section 2.4.2. Why not?

Answer

The directions of the errors for both of the test and reference samples were not random and independent.

It can be seen from Figure 15.5(a) that if the spectral bandwidth of the instrument, Δ_I, is *significantly* smaller than the analyte linewidth, Δ_S, then the effect on peak height can be negligible. It is normally accepted that the spectral bandwidth of the instrument should be not more than one tenth of the analyte linewidth.

In a spectrophotometer which uses a *diode-array* detector, the exit slit to the monochromator is replaced by the diode detector array itself (Section 9.3.2). The signal from each diode then passes directly into the software-processing unit. Thus, it is possible to convert the *transmittance* value measured by each diode *separately* into an *absorbance* value, *before* the signals from the 'active' diodes are aggregated. Figure 15.5(b) shows the effect of repeating the convolution of

Figure 15.5(a) against the *absorbance* spectrum, instead of the *transmittance* spectrum.

By comparing the two parts of Figure 15, it can be seen that when the convolution occurs in the absorbance spectrum the percentage error in peak height is *not dependent* on the *absorbance* of the sample.

DQ 15.7

How does Figure 15.5(b) show that it is possible to 'bunch' data points in an absorbance spectrum without loosing photometric linearity?

Answer

The important factor is that the percentage error is independent of the absorbance value, i.e. the percentage error is the same in both the reference sample and the test sample, and thus it cancels out in the ratio between them. Note that this is not true for a convolution in transmittance, as shown in DQ 15.5.

We find that wavelength bunching of the spectrum does *not* result in errors in calculating relative absorbances (e.g. measuring the concentration of a sample against a standard), *provided* that the bunching is carried out by using *absorbance* and not transmittance values. This allows the use of wavelength bunching as a viable method of reducing the signal-to-noise ratio without compromising comparative photometric accuracy. However, this can only be done with array detectors, and can not be achieved by simply opening the slit of a conventional spectrophotometer.

SAQ 15.2

Explain the difference between the following two processes and the reasons for using each of them:

(i) The 'bunching' of data points where groups of '*n*' data points are averaged together in order to provide a single data value, thus giving only one value for every '*n*' of the original data points.

(ii) 'Box-car' filtering where groups of '*n*' data points are averaged together by using a process of convolution, thus giving the same density of data values as the original set.

15.6 Deconvolution

We have identified the steps in the process of the convolution whereby two 'input' functions F_1 (true spectrum) and F_2 (convolute) produce an 'output' function

F_O (observed spectrum). The reverse process of *deconvolution* is also possible. Given the observed spectrum and the convolute F_2, it is possible to reconstruct the true spectrum.

However, the term deconvolution is also sometimes applied (incorrectly) to the process of disentangling separate lines or peaks that are *overlapping* to form a complex spectrum or chromatogram. Such a total spectrum is the result of simple *addition* of the separate spectra — it is *not true* convolution.

SAQ 15.3

Plot a graph of the apparent absorbance of a UV–visible spectral line against true absorbance from 0A to 4A. Assume that the spectral line has a linewidth of 20 nm, and the spectrophotometer has a spectral bandwidth of 8 nm.

Is the relationship that you obtain linear?

Note. A further question that relates to the topics covered in this chapter can be found in SAQ 7.5.

Summary

This chapter has aimed to give the reader a conceptual understanding of the process of convolution, and for this reason, the topic was introduced by using the practical example of the possible peak-broadening effect due to the volume of an HPLC detector cell.

It introduced the mathematical background for convolution and its use in software as a digital filter. The effects of the box-car and Savitsky–Golay convolutes on a Gaussian peak were examined with respect to possible distortion of the signal.

Other examples of convolution were also considered, in particular the effect of the triangular spectral bandwidth on the photometric response of a UV–visible spectrophotometer.

Reference

1. Savitzky, A. and Golay, M. J. E., *Anal. Chem.*, **36**, 1627–1639 (1964).

Chapter 16

Monochromators

Learning Objectives

- To identify the main performance characteristics of the monochromator.
- To quantify the trade-off between resolution and energy throughput.
- To appreciate the problem of unwanted 'orders' arising from the diffraction grating.
- To describe the construction and specific advantages of an echelle grating when used in inductively coupled plasma systems.

16.1 Introduction

Does a <u>monochromator</u> give only one wavelength?

We have seen in Section 5.6 that the electromagnetic radiation used in a spectrophotometric system is often required to have a narrow spectral bandwidth. This is an essential contribution to the *selectivity* of the instrument.

Optical filters (Section 13.7) can be used for wavelength selection, but they have a fixed wavelength transmission profile that can not be altered. A monochromator is an optical system that has a spectral transmittance profile that can be *adjusted* for both *peak wavelength* and, if necessary, *spectral bandwidth*.

Early monochromators used prisms as the dispersive (wavelength-selecting) element. However, prisms suffer from the fact that their dispersion is a function of wavelength, temperature, and the exact type of glass used. Automated wavelength selection by rotation of the prism is therefore difficult as it requires the use of a driving cam which is specifically contoured to match the dispersive properties of the glass used in its construction. However, prism systems are now reappearing in conjunction with detector arrays (see Section 18.4), where wavelength calibration can be achieved by using software without the need for mechanical *rotation* of the prism.

The diffraction grating is the main alternative to the prism, and when combined with a simple drive mechanism is able to produce a linear wavelength response. The development of high-quality gratings, which use the accuracy of an holographic pattern (see below, in Section 16.2.5), has now established them as the preferred choice for most applications.

Echelle gratings (a special form of diffraction grating) are used for specific applications, and we will discuss here a recent development that uses an echelle grating in conjunction with a prism to produce a high-resolution monochromator.

We have also discussed elsewhere (Section 7.2) the use of the Michelson interferometer as a wavelength-selective system, but this device produces a *sinusoidal* transmittance profile instead of the triangular profile of the typical monochromator.

DQ 16.1

What characteristics would be used to define the quality of a monochromator? (Hint: consider the UV–visible spectrophotometer.)

Answer

The characteristics required are spectral bandwidth, peak transmittance, wavelength accuracy and precison, stray light and energy throughput (f-no.).

16.2 Diffraction-Grating Monochromator

Why is it sometimes necessary to use a set of broad-band filters when the instrument already uses a monochromator?

A diffraction-grating monochromator uses the optical-*dispersion* properties of a *diffraction grating* to separate the wavelength components of light into a spectrum — see below in Section 16.2.1.

There are several different designs for monochromators, which use either flat or curved diffraction gratings, with the optical layout of a typical monochromator (Czerny–Turner design) being given in Figure 16.1.

Light from the 'entrance slit' falls on the first concave mirror, which produces a parallel beam of light which is then diffracted by the flat diffraction grating. The second concave mirror then focuses the diffracted light to produce an *image of the entrance slit* in the same plane as the *exit slit*.

The grating 'diffracts' light of different wavelengths through different angles. Hence, separate images of the entrance slit (for each wavelength) are produced at different positions along the plane of the exit slit. This spread of images forms a spectrum of the light.

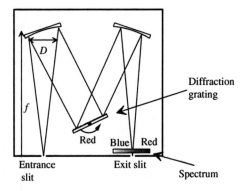

Figure 16.1 Schematic representation of the optical layout of a diffraction-grating mono-chromator (Czerny–Turner design); f is the focal length and D is the beam aperture.

A narrow part of this spectrum, centred about λ_0, will fall across the exit slit, and will then pass out through the latter. The spectrum is moved by rotating the diffraction grating. As the diffraction grating is turned (in the direction shown in Figure 16.1) the spectrum will move towards the left and a different wavelength will be selected to exit the monochromator.

16.2.1 Action of the Diffraction Grating

The diffraction grating has a reflective surface that has been etched with thousands of very fine parallel strips (lines). These strips are separated by a distance 'd', giving '$1/d$' lines per metre (see Figure 16.2).

The width of each strip is less than the wavelength of light, which means that the strips *do not act as simple mirrors*. When the incident light falls on each strip, the energy can be re-radiated at any angle. The effect is similar to sea waves spreading out in *arcs* after hitting the individual supporting posts of a pier.

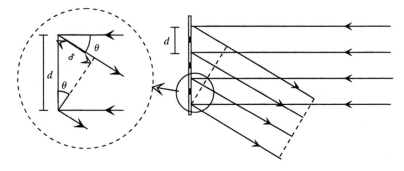

Figure 16.2 The action of a diffraction grating, showing its behaviour to incident light.

The light that leaves the grating is the sum of the components *diffracted* from each of the strips, and when the light components overlap each other, they will cause optical interference. It is possible to show that interference between the different components results in a concentration of energy of the reflected radiation into *particular directions*, where the latter depend on the wavelength of the radiation.

We will take a simplified case where the radiation hits the diffraction grating at 90°. At an angle of diffraction, θ, we can see from the enlarged cross-section shown in Figure 16.2 that components from adjacent strips travel an extra distance 'δ' (optical path difference (opd)) to arrive at the new wave front.

If, at a specific angle θ, the optical path difference, δ, between the light from adjacent strips is equal to an integer number of wavelengths ($n\lambda$), then the two adjacent components will still be in-phase (in step!) and will 'interfere' *constructively* (i.e. they will add up!). If we take into account the thousands of strips on a grating, it can be shown that *energy will only be diffracted at those angles* that satisfy the following equation:

$$\delta = n\lambda \tag{16.1}$$

At other arbitrary angles of diffraction, the wave components for all of the strips will have travelled distances that result in them arriving in varying phases. The overall result is *destructive* interference and *no energy* is diffracted in these arbitrary directions.

By using the equation for a right-angle triangle, we find the condition for 'reflection' is as follows:

$$\delta = d \times \sin \theta \tag{16.2}$$

Hence, light of wavelength λ is diffracted at an angle θ, which satisfies the following relationship:

$$n\lambda = d \times \sin \theta \tag{16.3}$$

The monochromator is therefore able to select light of a particular wavelength, λ_0, by rotating the grating to change the angle θ.

An important fact to notice is that the equation can be 'true' for different *integer* values of n. For each value of $n(= 1, 2, 3,$ etc.), a *separate spectrum* can be produced. The value of 'n' is said to be the *order* of the spectrum, e.g. first order, second order, etc.

Thus, for a given value of θ (i.e. position of the grating) it is possible for different values of the wavelength to satisfy equation (16.3). For example, a wavelength of 1200 nm in the first-order spectrum ($n = 1$) will occur at the same position (value of θ) as the following:

600 nm in the second-order spectrum: $n\lambda = 2 \times 600 = 1200 = d \sin\theta$;
400 nm in the third-order spectrum: $n\lambda = 3 \times 400 = 1200 = d \sin\theta$;
300 nm in the fourth-order spectrum: $n\lambda = 4 \times 300 = 1200 = d \sin\theta$;

and so on.

Light of all of these wavelengths will be able to pass out through the exit slit for the same diffraction angle.

DQ 16.2

The diffraction grating in a (flame) atomic absorption spectrophotometer has 1800 lines/mm and is used in the first-order mode. Calculate the wavelengths of light, due to unwanted orders, that could be transmitted when the monochromator is set to 900 nm

Answer

The additional wavelengths will be given by $n\lambda = 1 \times 900$, *where n is an integer. This gives 450, 300, and 225 nm. The fifth-order value, i.e. 180 nm, would be outside of the normal UV–visible range of such an instrument.*

It is often necessary to absorb the unwanted orders by using a separate broad-band filter in line with the monochromator. Figure 16.3 shows that the monochromator transmits very narrow bands of radiation at several related wavelengths which correspond to the different orders. The filter, however, transmits a broad range of wavelengths. By combining the transmission of these two units, only the desired wavelength will be transmitted.

It is still possible to change the monochromator wavelength within the range of such a filter, although it is often necessary to have a number of different filters that can be switched into place for different wavelength ranges.

The other problem posed by the existence of unwanted orders is that they take energy *away* from the spectrum being used. However, the diffraction grating can be designed to reflect most of its energy into a particular order by using a process known as 'blazing'. In this, each strip is etched at an appropriate angle that enhances the light in the direction of simple reflection from the strip surface. Blazing is most effective at a particular wavelength, and this is normally chosen to be roughly in the middle of the designed operational range for the monochromator.

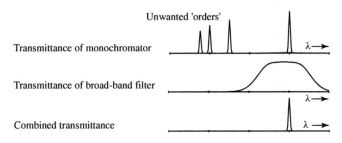

Figure 16.3 Demonstration of the effect of a broad-band filter when used in combination with a monochromator — 'order sorting'.

16.2.2 Spectral Bandwidth

The angular spread of wavelengths (*angular dispersion*) from the diffraction grating is given by differentiating equation (16.3) to obtain the following:

$$\frac{d\theta}{d\lambda} = \frac{n}{d\cos\theta} \tag{16.4}$$

If the focal length of the monochromator is f, then the angular spread of wavelengths becomes a linear spread of wavelengths (*linear dispersion*) at the exit slit, as given by the following:

$$\frac{dx}{d\lambda} = f\frac{d\theta}{d\lambda} = \frac{fn}{d\cos\theta} \tag{16.5}$$

where x is the linear distance along the spectrum. Note that a large value of $dx/d\lambda$ means that the spectrum is spread over a larger distance.

If the exit slit has a width W, then the range of wavelengths passing through the slit, i.e. the *spectral bandwidth*, is given by the following:

$$\Delta\lambda_M = \frac{d\lambda}{dx} \times W = \frac{d\cos\theta}{fn} \times W \tag{16.6}$$

It can be seen that the spectral bandwidth can be *reduced* (i.e. an increased resolution) by the following:

- decreasing the slit width, i.e. making W smaller;
- using a monochromator with a longer focal length, i.e. increasing f;
- using a diffraction grating with more lines per metre, i.e. reducing d;
- using a higher-order spectrum, i.e. a larger n.

The resolving power of the monochromator is given by the following expression:

$$R_P = \frac{\lambda_0}{\Delta\lambda_M} \tag{16.7}$$

DQ 16.3

Estimate (taking $\cos\theta = 0.5$) in the *first-order* mode, the actual width of the slits in a monochromator with 1800 lines/mm, a focal length of 0.3 m, and a spectral bandwidth of 0.1 nm.

Answer

The distance between each line in the grating, d, is $1/1800 \times 10^3$ m = 5.556×10^{-7} m. The slit width is found by simple substitution into equation (16.6), as follows:

$$W = \Delta\lambda_M fn/(d\cos\theta) = 0.1 \times 10^{-9} \times 0.3 \times 1/(5.556 \times 10^{-7} \times 0.5)$$

$$= 1.08 \times 10^{-4} \text{ m, i.e. the slit is approximately 0.1 mm wide.}$$

SAQ 16.1

A diffraction-grating monochromator has a linear dispersion of 0.5×10^6 at a wavelength of 500 nm. If the slit width is 0.5 mm, calculate (i) its spectral bandwidth, and (ii) its resolving power.

16.2.3 Transmission Spectral Profile

The monochromator normally has a triangular transmission profile. This is illustrated in Figure 16.4 for a monochromator that gives maximum transmittance when it is adjusted to λ_0. This figure shows the various positions of the image of the entrance slit in relation to the exit slit for different wavelengths (a), along with the corresponding transmittance diagram (b) which shows the proportion of light transmitted for each wavelength.

For the optimum setting, λ_0, the image of the entrance slit exactly overlaps the exit slit, and maximum energy is transmitted ($T = T_{max}$). Light of wavelength λ_1 is of a shorter wavelength and the image is too far to the left to allow energy to pass through the slit ($T = 0$). Similarly, light of longer wavelength, λ_4, has an image outside of the exit slit and again $T = 0$.

For light of wavelengths λ_2 (half-way between λ_1 and λ_0), the image *half* overlaps the exit slit and *half* the energy will be transmitted, i.e. $T = T_{max}/2$. The same reasoning applies to λ_3. The *linear* increase in overlap and proportion of light transmitted thus gives the *triangular* shape to the transmission profile.

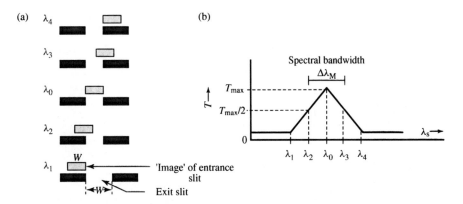

Figure 16.4 Generation of a triangular spectral bandwidth, showing (a) the various positions of the entrance slit in relation to the exit slit at different wavelengths, and (b) the corresponding transmittance profile.

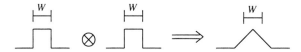

Figure 16.5 Illustration of the convolution of two rectangular functions to give a triangular function.

The widths of the entrance and exit slits are the same so that, in perfect adjustment, the image of the entrance slit exactly fills the exit slit. The relationship between $\Delta\lambda_M$ and the width W is given above in equation (16.6).

The process described above is mathematically the same as the convolution of two identical rectangular functions (Section 15.5) to produce a triangular function (Figure 16.5).

16.2.4 Energy Throughput

What is the problem in trying to improve resolution by making the slit width very, very, small?

Light is focused on the *entrance* slit of the monochromator in order to produce the internal spectrum. The *total energy* in this spectrum is therefore proportional to the width of the *entrance* slit, W. The *proportion* of the final spectrum which passes through the exit slit is then also proportional to the width of the *exit* slit (also W). Thus, the total energy throughput is proportional to $W \times W = W^2$.

If the slits are narrowed to improve resolution (a smaller spectral bandwidth), the available energy drops as the *square* of the spectral bandwidth. A very narrow bandwidth means very low energy levels reaching the detector, and consequently more sophisticated detection systems will be needed.

DQ 16.4

Reducing the slit width to improve resolution also reduces the available energy. Why is this a problem? For example, does it:

(a) reduce the apparent value of the absorbance;
(b) reduce the signal-to-noise ratio in the measurement;
(c) increase the time taken to make the measurement?

Answer

*The **calibration** of light energy by using a reference sample would ensure that there is no systematic error in the absorbance of the test sample (a, false). However, the reduction in light energy would reduce the magnitude of the analytical signal in comparison to the magnitude of the noise created in the system, i.e. the S/N ratio would be reduced (b, true). One* *method of re-establishing the original S/N ratio would be to average the signal over a longer time ((c) could therefore also be true).*

The f-number of the optical system is also an important factor in the light-transmission efficiency. This is given by the ratio of the focal length, f, to the beam aperture, D (see Figure 16.1), as follows:

$$f\text{-number} = f/D \tag{16.8}$$

The same optical characteristic is used for cameras, where D is the aperture of the lens, and the energy transmitted is inversely proportional to the square of the f-number. However, low f-numbers require better quality optical systems due to the larger angles with which the beam diverts from the axis.

DQ 16.5

In a (flame) atomic absorption spectrophotometer, which of the following statements are true in relation to the signal-to-noise ratio in the measurement? Explain your reasoning.

(a) A low f-number is required to maximize energy throughput.

(b) A high f-number is required because its low beam divergence (see Figure 16.1) allows good alignment with the flame.

Answer

A low f-number gives greater signal energy and would therefore increase the S/N ratio. However, a low f-number gives a greater beam divergence, which makes it difficult to concentrate the beam along the central part of the flame. The outer parts of the flame generate greater background emission 'noise', which will reduce the S/N noise, thus giving particular problems in trace analysis. Hence, a balance in design must be made for optimum performance.

SAQ 16.2

The peak transmittance of a monochromator with a spectral bandwidth of 4 nm is 80%. If the slits are narrowed to produce a bandwidth of 2 nm, calculate the following:

(i) the change in total energy transmitted;
(ii) the new peak transmittance.

16.2.5 Stray Light

Most stray light originates within the monochromator, and can be due to a number of causes. Examples of those include the following:

- light in unwanted 'orders' which has not been absorbed by additional filters;
- light which has been scattered, diffracted or reflected from imperfections in the optical system;
- light diffracted from component edges.

The stray light figure, S_L, is given by the following:

$$S_L = \frac{\text{Intensity of } \textit{unwanted} \text{ radiation}}{\text{Intenstity of } \textit{wanted} \text{ radiation}} = \frac{I_S}{I_0(\lambda)} \qquad (16.9)$$

where I_S is the intensity of the stray light.

One source of stray light is an uneven ruling of the lines on the diffraction grating. Diffraction gratings were initially produced mechanically, i.e. by using a diamond-tipped tool to cut grooves in a master grating. However, it was extremely difficult to ensure uniformity of spacing over several thousands of lines, and this non-uniformity resulted in 'ghost' lines of the wrong wavelength. Modern gratings are produced by a photo-etching process, using an optically produced *holographic* image. These holographic gratings have far greater uniformity across the width of the grating and thus give less stray light.

16.3 Echelle-Grating Monochromator

An Echelle grating uses 'reflections' from the faces of a *stepped* grating (see Figure 16.6), with $\theta \approx 90°$ ($\cos \theta \approx 0$) and in high-order (large n) spectra. This gives the following:

- a high linear dispersion (equation (16.5));
- a narrow bandwidth (equation (16.6));
- overlapping orders.

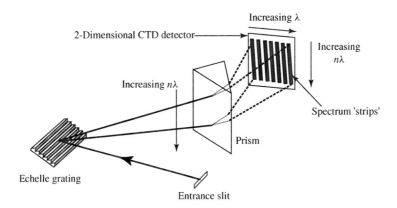

Figure 16.6 Schematic representation of an echelle-grating monochromator.

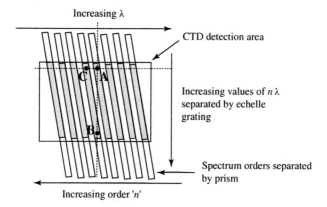

Figure 16.7 Two-dimensional spectral output obtained from an echelle-grating mono-chromator.

The echelle grating has specific advantages for atomic absorption spectrophotometry, in that the narrow spectral bandwidth is compatible with the narrow linewidth of 'atomic absorption' and provides good rejection of the background radiation.

For inductively coupled plasma systems, which have a large number of emission lines, the overlapping orders present a problem. However, this can be resolved by using a prism with its direction of dispersion perpendicular to that of the grating. In Figure 16.6, the echelle grating produces a single 'vertical' spectrum with overlapping orders, but the prism ensures that the different orders are separated 'horizontally', thus giving a series of (nearly) vertical spectra, as shown in Figure 16.7.

Referring to the resultant two-dimensional spectral map shown in Figure 16.7, we can note the following features:

- points C and B are in the same spectral order (n), but the wavelength at B is greater than that at C;

- points C and A have the same value of ($n \times \lambda$) because they have been diffracted through the same vertical angle by the grating;

- points A and B have the same wavelength because they have been refracted through the same horizontal angle by the prism.

SAQ 16.3

The layout of a spectrophotometer includes a 'double monochromator', which appears as one monochromator following another. Both monochromators must be set to the same wavelength, or otherwise no radiation could pass through them both. What then is the advantage of selecting the same wavelength twice?

SAQ 16.4

Figure 16.7 shows the two-dimensional output from the echelle monchromator of an (ICP) atomic emission spectrophotometer, which uses a prism for orthogonal dispersion.

(i) If the wavelength at point A is 324 nm in the 50th order, what are the wavelengths at points B and C?

(ii) Which part of the two-dimensional spectral output has the largest dispersion — at long (near-IR) or short (UV) wavelengths?

Note. Further questions that relate to the topics covered in this chapter can be found in SAQs 12.2, 15.1, 18.2 and 18.3, and DQs 3.6, 5.1 and 5.9.

Summary

This chapter has concentrated on the performance characteristics of the diffraction-grating monochromator, in particular the spectral bandwidth, energy throughput and unwanted orders.

It also looked at the use of the echelle grating in providing the very high resolution that is useful in modern atomic spectroscopy.

Chapter 17

Radiation Sources

Learning Objectives

- To differentiate between a photon emitter and a thermal emitter of radiation.
- To calculate the energies in electronvolts which are equivalent to the energies of photons of given wavelengths.
- To identify characteristics associated with different types of light sources.
- To describe the particular characteristics of laser light which are appropriate to instrumental applications.

17.1 Introduction

How does the radiation spectrum from a filament lamp differ from that of a discharge tube?

Photons of electromagnetic radiation are emitted when a system moves from a higher-energy state to a lower-energy state. The frequency, v, and wavelength, λ, of the photon emitted is determined exactly by its energy, E, as follows:

$$v = E/h; \lambda = c/v = ch/E \qquad (17.1)$$

where c is the speed of light and h is the Planck constant.

In atomic emission, the electrons move between *atomic* energy levels. If the atom is free of any interaction with other atoms, the amount of energy liberated can be very precise, and all of the photons share a very clearly defined wavelength. This is the characteristic sharp 'line' emission from an electrical discharge in a low-pressure gas. A very good example of this is the hollow-cathode lamp (see Section 17.3.3) used in atomic absorption spectrophotometry.

However, the Doppler effect will modify the wavelength due to the *speed* of the atom towards, or away from, the observer. At *higher temperatures*, therefore, the emitted line will be broadened due to the increased random thermal motion of the atoms (Doppler broadening).

In addition, if the atoms interact *with one another*, the energy of interaction will broaden the possible energies of transition, thus causing a broadening of the spectral lines. If the pressure in a gas discharge is increased, the atoms interact more, thus resulting in pressure broadening.

In the case of a hot solid object, the interactions between atoms are so strong that the emitted spectrum becomes continuous. The emitted radiation is no longer characteristic of specific transitions (quantum emission) but approaches the thermodynamic concept of a perfect 'thermal emitter' (thermal emission).

We can therefore differentiate between two *types* of emission process, as follows:

- Quantum emission — characteristic of specific energy transitions;
- Thermal emission — limited by thermodynamic considerations.

17.2 Thermal Emitters

How is it possible to tell the temperature of a furnace just by looking at the colour of the light being emitted?

It is a familiar fact that as its temperature increases, an object first glows a dull red, turns to red–orange, and then to a 'white' heat when sufficiently hot. The radiation from even hotter sources (e.g. a plasma or a high-voltage electric discharge) contains wavelengths extending into the blue and UV region of the spectrum.

The emission of radiation from an incandescent solid can be approximated to the emission of radiation from a 'perfect emitter'. Thermodynamics allows us to calculate this theoretical spectral emission at a given wavelength and temperature. Figure 17.1 shows the relative thermal emission at temperatures of 2600, 3000 and 3400 K. We see that, as the source becomes hotter, the output increases rapidly and the wavelength of maximum emission moves towards the blue end of the spectrum.

The principle features of such spectra are as follows:

(a) The total energy emitted (across all wavelengths), given by:

$$E_0 = \sigma A T^4 \tag{17.2}$$

where σ is the Stefan constant (5.67×10^{-8} W m^{-2} K^{-4}), A is the surface area (m^2), and T is the temperature (K).

(b) The wavelength of maximum emission, given by:

$$\lambda_{\text{max}} = b/T \tag{17.3}$$

where b is the Wien constant (2.898×10^{-3} m K).

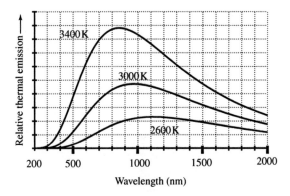

Figure 17.1 Radiation output profiles obtained from a perfect emitter at temperatures of 2600, 3000 and 3400 K.

Figure 17.1 shows the emission, $E_{0,\lambda}$, at a given wavelength for a theoretically 'perfect' emitter. The emission, E_λ, from a *real* surface will be a fraction, ε_λ, of the theoretical value, thus giving the following:

$$E_\lambda = \varepsilon_\lambda \times E_{0,\lambda} \qquad (17.4)$$

where ε_λ is the *emissivity* of the surface at a wavelength λ.

Thermodynamics can prove that the emissivity of an object is equal to its absorptivity (the fraction of light absorbed). As a good absorber is also a good emitter, and the best absorber of radiation is a 'hole' which 'absorbs' all radiation, we therefore reach the surprising conclusion that the best *emitter* of radiation is also a 'hole'.

In practice, we can make an *emitting* 'hole' by constructing a cavity inside a heated block, and opening a small hole in the side of the block to allow light to enter or leave. Any radiation entering the cavity is almost entirely absorbed by baffles and blackened surfaces. As the hole is then an almost perfect *absorber*, the hole will also act as an almost perfect *emitter* for the radiation produced when the block is heated to a sufficient temperature.

Cavity sources are used in some instruments for IR wavelengths. However, the most common type of thermal-emission source is still the heated filament.

DQ 17.1

Which of the following is the best temperature to use for a source suitable for IR radiation at about 1000 cm^{-1}:

(a) 300 K;
(b) 1300 K;
(c) 3000 K?

Answer

*Radiation of wavenumber 1000 cm⁻¹ has a wavelength of 1.0 × 10⁻⁵ m. By using equation (17.3), we find that for **peak emission** to be at this wavelength, the temperature should be 290 K. However, this is no higher than normal room temperature, and all objects in the instrument, including the box itself, will be at about the same temperature and thus emitting radiation! A source of radiation must have a higher emission that its surroundings, and so a **higher temperature** is required. A temperature of 3000 K would, of the three options, give the most radiation at 1000 cm⁻¹, but it would **also** give an excessive amount of energy at wavelengths outside of those required. A compromise temperature is therefore chosen, with practical sources operating at about 1300 K.*

SAQ 17.1

Why do light bulbs always seem to 'blow' when you need them most, i.e. when you first switch them on?

17.2.1 Tungsten–Halogen Lamp

How 'white' is the light from a tungsten lamp?

The emission of visible light from a heated filament increases dramatically with an increase in filament temperature (equation (17.2)). However, in normal operation a slight increase in temperature rapidly increases the evaporation of atoms from the surface of the filament, thereby reducing the life of the lamp. The design of such lamps is therefore a careful balance between emission efficiency and lifetime. In some spectrophotometers, the lamp is sometimes run at a slightly lower temperature in order to extend its life.

In the *tungsten–halogen* lamp, the filament is run at a higher temperature than the normal tungsten lamp. This takes advantage of the Wien and Stefan relationships to provide a source with a maximum wavelength nearer the centre of the visible spectrum (whiter) and with a greater energy output (brighter) — see Figure 17.1. The problem of increased atom evaporation from the high-temperature filament is counteracted by introducing a halogen gas into the lamp envelope.

When the tungsten atoms evaporate, they bond with halogen atoms to produce volatile molecules that prevent the tungsten being deposited and lost on the inside of the envelope. These molecules only dissociate when they come into contact with the hot filament, at which time the tungsten is 'put back' on to the filament. The halogen atoms are acting as scavengers — collecting evaporated tungsten atoms and returning them to the filament.

17.3 Quantum Emitters

Why are electronvolts important?

Most quantum sources of radiation use *electrical energy* to put the atom into the required higher-energy state. If an electron is accelerated by a voltage difference of V volts, then the energy gained by that electron is V electronvolts (eV), where 1 eV is equivalent to 1.6×10^{-19} joules (J) (e ($= -1.6 \times 10^{-19}$ coulombs (C)) is the charge on the electron).

In an *electrical gas discharge*, a voltage is applied between two electrodes in a gas (or vapour). Free electrons and ions are accelerated by the electric field and collide with the atoms, thus causing further excitation and ionization. Much of the energy, which is released when the atoms return to their ground states, is emitted as photons of electromagnetic radiation.

A process of emission can also take place within *semiconductor* materials. Electrons travelling through a p-n junction in a diode can 'drop' from one energy level to another as the voltage changes, with the emission of a photon of energy. Light-emitting diodes (LEDs) are common sources (see Section 17.3.5 below) that make use of this process.

> ## DQ 17.2
>
> Calculate the minimum excitation voltage required before a source can generate photons of wavelength 500 nm.
>
> ### *Answer*
>
> *The photon energy for 500 nm = hc/λ = (6.6 × 10^{-34} × 3.0 × 10^8)/ (500 × 10^{-9}) = 3.96 × 10^{-19} J. This energy is equivalent to 2.5 eV. Hence, each electron must be accelerated by a voltage of at least 2.5 V in order to acquire enough energy to generate such a photon. Note, for example, that the voltage applied across 'blue' LEDs is greater that it is for 'red' LEDs.*

17.3.1 Deuterium Lamp

The deuterium lamp is the standard lamp for producing UV radiation in analytical instruments. The voltage applied is typically of the order of 100 V, which gives the electrons plenty of energy to excite the deuterium atoms in a low-pressure gas to emit photons across the full UV range.

This type of lamp is basically a quantum emitter, although line broadening results in a continuous overlap of lines in the UV range. The visible emission line at 656.1 nm is often used for calibration of the monochromator.

17.3.2 Xenon Lamp

The strong atomic interaction in a high-pressure xenon discharge produces a continuous source of radiation which covers both the UV and visible range. The discharge is normally pulsed on for short periods, with a frequency that determines the average intensity of the light from the source, and also (inversely) its lifetime.

This lamp suffers from the problem of achieving repeatability of the light output between pulses. However, it is possible to use a secondary detector, which monitors the intensity of each pulse, and thus enables the instrument to compensate for any variations.

A remaining problem for this type of lamp is that the distribution of energy *between* wavelengths may change between pulses, even though the total energy output remains unchanged.

SAQ 17.2

What lamp sources should be used in an HPLC UV detector?

17.3.3 Hollow-Cathode and Electrodeless Discharge Lamps

Hollow-cathode lamps (HCLs) and electrodeless discharge lamps (EDLs) both provide atomic emissions from free metal atoms by using the heat of a gaseous discharge (e.g. in neon) to vaporize metal atoms into the discharge. HCLs use conventional electrodes, while the discharge in the EDL is induced by a radiofrequency or microwave field. Typically, each lamp is specific to a particular metal. However, some lamps use two, three or more metals, although the emission from each of these is reduced.

Due to the low vapour pressure of the metal atoms, there is little atomic interaction, thus giving an output spectrum with very narrow lines.

DQ 17.3

Is the light output intensity of an HCL controlled by:

(a) the applied voltage across the lamp;
(b) the current through the lamp?

Answer

The major factor is the current. The lamp can be kept on 'stand-by' with a minimum current which just maintains the discharge. A higher current will dissipate more power into the lamp, thus giving a greater light output. However, an increased current will also reduce the lifetime of the lamp.

As can be seen from the above discussion question, the radiation output of an HCL increases with increasing current. However, a large current causes excessive evaporation of metal atoms, which can then reabsorb the emitted radiation. This reduces the output at the central-peak wavelength while leaving the output in two wings on either side — see the discussion of the Smith–Hieftje method in Section 6.3.3.

17.3.4 Lasers

Lasers have TWO useful characteristics as sources of light — they give a very intense narrow beam AND ......?

Laser is an acronym for <u>L</u>ight <u>A</u>mplification by <u>S</u>timulated <u>E</u>mission of <u>R</u>adiation. The laser process can be made to occur either in a gas discharge or in a light-emitting diode.

An essential feature of the emission process in this case is that some of the emitted radiation must be reflected back through the system to stimulate further transitions. A form of resonance is established as the light which is reflected back maintains the coherence and power of the stimulated emission. A poor analogy of this is the 'howl' that builds up rapidly when the microphone of a public-address system is placed too close to the loudspeaker, with the system then reamplifying the feedback signal.

The laser resonance ensures that all of the separate atomic 'sources' of radiation are emitting *in phase* with one another. This provides the following three important features of laser light:

(i) *Intensity of emission* — the *co-ordinated* emission, from all of the 'sources', provides for much greater power output (constructive interference) than could be obtained if they were all emitting randomly. However precautions must be taken with laser light due to the possibility of eye damage.

(ii) *Narrow spectral bandwidth* — the resonance in the system ensures that the laser light has a very well-defined wavelength, with a *very narrow spectral bandwidth*.

(iii) The resonance also ensures that the wave is *coherent* over a much larger distance than would be the case with random emission from the separate sources. NB — a wave that is not coherent has frequent random changes in phase.

DQ 17.4

A laser is used in an FTIR spectrophotometer. Is this because it provides:

(a) intense light;
(b) coherent light?

Answer

The laser in an FTIR spectrophotometer is used to establish a position reference for the moving mirror. This requires a wave pattern which

*remains coherent over the length of mirror travel — only a laser light
can provide this coherence. Intensity is not important for this function.*

17.3.5 Solid-State Sources

The use of light emitting diodes (LEDs) as indicator lights is now a very common
feature in all types of equipment. However, they also have a role as simple
radiation 'sources' in situations where a lower power output is acceptable. LED
laser sources are also very common, e.g. in reading CDs.

The wavelength range for LED emission is limited by the availability of materials that have the required energy transitions. Common semiconductor materials
provide energy transitions of the order of 1 eV, which gives a wavelength of about
1250 nm (in the near IR). The production of materials that can emit shorter wavelengths has required considerable research and development. Red LEDs became
available first, followed by a gradual progression to yellow, green and, more
recently, blue LEDs.

The use of solid-state sources and detectors, coupled with optical fibres to
channel the radiation, is now enabling the development of sophisticated sensor
devices that incorporate miniature spectrophotometric measurement systems.

SAQ 17.3

The output from a tungsten spectrophotometer lamp deteriorates through age.
Which of the following statements are likely to be true?

(a) The wavelength setting of the instrument requires recalibration   (True/False?).

(b) The photometric accuracy at 650 nm is likely to become significantly worse
(True/False?).

(c) High-frequency noise will appear in the output from the lamp   (True/False?).

(d) It may no longer be possible to set 0*A* at a wavelength of 350 nm   (True/False?).

(e) The output signal from the instrument will become noisier   (True/False?).

(f) The spectral bandwidth of the instrument will increase   (True/False?).

Note. Further questions that relate to the topics covered in this chapter can be
found in SAQ 18.2, and DQs 2.4, 5.3 and 9.9.

Summary

This chapter has differentiated between the photon emission of radiation and
thermal emission, and identified particular characteristics appropriate to the
different mechanisms.

It discussed different practical sources for UV, visible and IR radiation,
including laser and solid-state sources.

Chapter 18

Radiation Detectors

Learning Objectives

- To explain why noise is generally a greater problem for the detection of IR photons than it is for UV–visible photons.
- To identify the relative advantages of different types of detectors.
- To differentiate between the two main types of charge-transfer detectors.

18.1 Introduction

Modern detectors employ both the earliest (vacuum tube) and the latest (charge-coupled device) technologies.

The performance characteristics of all detectors include the following:

- *Generic* characteristics (Section 3.2) — responsivity, linearity, dynamic range, noise, drift, etc.;
- *Specific* characteristics which depend on the detection processes being used.

There are two main mechanisms for detecting electromagnetic radiation, namely photon detection and thermal detection.

In photon detection, the radiation photon interacts directly with the material of the detector, and liberates an electron from a bound state into a *mobile* state. This may involve the total release of an electron from the detector surface (the photoelectric effect), or the excitation of an electron into a higher energy state within the detector material. The mobile state of the electron can be recorded as a flow of charge, i.e. as an electric current.

The *quantum efficiency* of a photon detector is the ratio of the number of 'released' electrons to the number of photons.

In thermal detection, the photon energy in the radiation is initially converted into *heat*, which causes a slight increase in the temperature of the detector. A secondary process is then used to record this change of temperature.

The major limitations of the thermal detection process are as follows:

(i) the signal-to-noise ratios are lower because the thermal noise is significant in comparison to the temperature fluctuations caused by the signal;

(ii) the speed of response of the detector is restricted by the thermal inertia of the detector material.

A disadvantage of the photon process is that the quantum efficiency is a function of both the photon wavelength and the detector material. Different materials have different characteristic spectral responses and operate over different wavelength ranges. Conversely, thermal-detection systems have the advantage that, provided the target is a good absorber, the spectral response of the detector is independent of the material, and will operate across all of the wavelengths for which it remains a good absorber.

A vital characteristic of any detector is the level of noise produced in the device itself. The design of the device must seek to differentiate between the response to the analytical signal and the production of a noise signal, i.e. it should have a high S/N ratio.

DQ 18.1

(i) Calculate the energies, in eV, of photons in the UV, visible, and mid-IR regions of the spectrum. Choose wavelengths of 250, 500 and 5000 nm, respectively.

(ii) Estimate the energy, in eV, which is equivalent to the thermal energy, kT, at room temperature.

Answer

(i) *For $\lambda = 250$ nm, the photon energy* $E(= hc/\lambda)$ $(= 6.6 \times 10^{-34} \times 3.0 \times 10^8)/(250 \times 10^{-9}) = 7.92 \times 10^{-19} J = (7.92 \times 10^{-19})/(1.6 \times 10^{-19})$ *eV* $= 4.95$ *eV. Similarly, the photon energies for $\lambda = 500$ and 5000 nm are 2.48 and 0.25 eV, respectively.*

(ii) *At room temperature (25°C),* $kT = 1.38 \times 10^{-23} \times 298 = 4.11 \times 10^{-21} J = 0.026$ *eV.*

It can be seen from the above discussion question that there is a considerable difference in energy between UV–visible quanta ($h\nu > 2$ eV) and thermal quanta ($kT \approx 0.026$ eV). Hence, UV–visible detectors have good signal-to-noise ratios. However, DQ 18.1 also shows that a detector that responds to mid-IR photons is likely to be more responsive to thermal quanta. It is therefore more difficult to develop efficient photon detectors for the mid-IR region, and those that have been developed (see Section 18.5 below) usually require cooling to reduce the thermal noise.

Due to the quantized nature of the signal, photon-detection systems are inherently sensitive to shot noise (Section 14.4.2). This is often the limiting factor for the signal-to-noise ratios in UV–visible detectors.

In addition for photon detectors, other sources of energy (e.g. natural radioactivity and thermal energy) may excite the electrons into the mobile state, thus giving a false signal. The latter is called a 'dark current' because it will exist even if no light is falling on the detector.

It is possible to allow for the *constant* error signal due to the dark current by adding an offset to the detector amplifier. However, the dark current itself has inherent *random* shot noise, which cannot be counteracted, and this will be an ultimate limit to the detectivity of the detector. The design of the detector must seek to minimize the dark current and the associated noise, while still trying to maximize its responsivity to the photons of the analytical signal.

SAQ 18.1
Why are some radiation detectors cooled while others are not?

18.2 Phototube and Photomultiplier

Electronic valves have been replaced by integrated-circuit devices — why are vacuum-tube detectors still used?

The phototube (photocell) is illustrated in Figure 18.1(a). When the photon strikes the cathode, the photon energy, E, is sufficient to release an electron into the vacuum. This electron is attracted to the anode by the positive voltage and will then pass through the external circuit as an electric signal current, I_S.

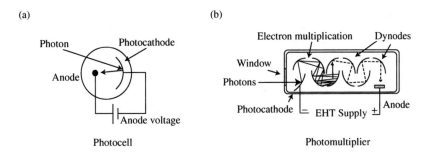

Figure 18.1 Schematic representations of (a) a phototube (photocell) and (b) a photomultiplier tube.

The photomultiplier tube, shown in Figure 18.1(b), has a 'front end' similar to the basic phototube. However, in this system a series of 'dynodes' are positioned between the cathode and anode, with a potential difference of about 100 V between adjacent dynodes. The total extra-high tension (EHT) voltage across a tube with 10 dynodes is therefore about 1000 V.

When the electron is released from the cathode, it is attracted to the first dynode by the voltage difference of 100 V, thus acquiring a kinetic energy of 100 eV. With this energy, the first electron 'releases' a number of new electrons, g, which are then accelerated to the second dynode. At the second dynode, *each* of the 'g' electrons can release a *further* 'g' electrons, thus giving a total of g^2 electrons. After 'n' amplification stages, the increase in the number of electrons ('G') leaving the photomultiplier tube (PMT) is given by the following:

$$G = g^n \qquad (18.1)$$

With such a system, it is quite normal to obtain gain factors as high as a million.

Clearly, this internal amplification makes the PMT a much more sensitive detector than the basic phototube. However, the PMT is also more sensitive to the dark current, which is also amplified by the same factor.

Tungsten is used as the normal material for the electrodes in most vacuum tubes, due to its favourable mechanical properties. However, the energy (work function) required to release a single electron from the surface of tungsten is 4 eV. This is no problem at the *dynodes* because the arriving electrons have energies of 100 eV. However, only photons with wavelengths *less than 310 nm* have energies greater then 4 eV, and it would therefore appear that only UV photons could liberate the initial electron from the cathode. Fortunately, it is possible to coat the cathode with another material (typically an alkaline-earth metal) which has a *low work function*. This treatment then allows this type of detector to be used in both the visible and near-IR regions, as well as the UV region.

Although coated cathodes can operate outside of the UV range, their spectral responsivity depends on the particular coating material being used. This limits their operational wavelength ranges, and it is sometimes necessary to use more that one detector for optimum sensitivity over a wide operational range, e.g. changing to a 'red' PMT is required by some spectrophotometers for wavelengths greater than 600 nm.

DQ 18.2

Increasing the EHT voltage on the photomultiplier tube increases the voltage between each pair of dynodes.

Which of the following statements is (are) correct:

(a) the responsivity of the PMT will increase;
(b) the PMT will respond to a wider range of wavelengths;

(c) the dark current will remain the same;

(d) the noise due to the dark current will increase?

Answer

The increased voltage gives the electrons greater energy when they strike each dynode, thus giving increased values of the responsivity factors, 'g' and 'G' (a, true). The increased gain will also increase the dark current at the output of the PMT (c, false), and it will then also increase the noise (Section 14.4.2) due to the dark current (d, true). The wavelength range depends primarily on the coating material of the cathode, but not on the applied voltage (b, false).

SAQ 18.2

Estimate an 'order of magnitude' value for the current obtained from a vacuum photocell in a single-beam spectrophotometer where:

- the 20 W tungsten lamp source emits approximately 10% of its energy in the visible spectrum between 350 and 650 nm;
- about 15% of the total radiant output from the lamp enters a 50% efficient monochromator, which has a bandwidth of 2 nm and is set to about 500 nm;
- the sample has an absorbance of 2A;
- the quantum efficiency of the photocell is 8%.

18.3 Photodiode

In solid-state detectors, the detection process occurs when an incoming photon excites an electron into a *mobile state* within a semiconductor material. This material is normally silicon, although other semiconductors have been developed for lower-energy photons.

In the photodiode, the 'liberation' of the electron (together with an associated 'hole') allows current to flow across a p-n junction, which has been 'reverse-biased' in order to prevent the *normal* flow of current. This *reverse* current will be proportional to the number of photons.

The spectral response range depends on the energy of transitions within the material. Enhanced UV-response silicon detectors have now been developed in order to increase the response into the ultraviolet region, thus giving a wide response range from the UV through to the near-IR.

The advantage of photodiodes is that they are easy to produce (by using modern semiconductor technology), they do not require the high voltages necessary for 'vacuum-tube' devices, and they are easily compatible with modern electronics. The disadvantage is that they are noisier than the vacuum-tube method of detection.

18.3.1 Photodiode Array

By using the modern fabrication techniques of microelectronics, it is now possible to produce a linear (one-dimensional) array of several hundred (or more) photodiodes set side-by-side on a single integrated circuit (IC), or 'chip'. Each diode is capable of recording the light intensity at one point along the line, and together they provide a linear profile of the light variation along the array.

However, the problem arises of how to amplify and record the hundreds of individual signals that are produced. It is not feasible to connect a separate amplifier to each diode, so a multiplex method is used, which 'reads-out' the signals one by one, and then feeds them *sequentially* to a single amplifier.

Each diode, D_x, is connected, in reverse bias, to a separate capacitor C_x, which is initially given a charge, Q_x^0. When light falls on the diode over a fixed period, T_E (exposure time or integration time), the diode will conduct current and partially discharge the capacitor. The drop in capacitor charge, ΔQ_x, will be proportional to the product of the the exposure time and the light intensity, I_x, falling on the particular diode, as follows:

$$\Delta Q_x = R_x \times I_x \times T_E \qquad (18.2)$$

where R_x is a factor which represents the responsivity of the particular diode.

At the end of the exposure time, each capacitor is holding a charge, $Q_x^0 - \Delta Q_x$. Each of these charge 'packets' is then switched sequentially to an amplifier system, which converts it into a voltage signal.

The output of the photodiode array (PDA) is a histogram profile, along the array, of the charge leaked by each photodiode. This mirrors the variation of light intensity across the array. It is, of course, necessary to *calibrate* each diode for *responsivity* and *offset*, but this is carried out in the software data-processing unit.

The cycle of detection has three main stages, as follows:

(i) initialization;
(ii) accumulation of charge at each pixel — integration time;
(iii) read-out of signals.

In comparison to the photomultiplier, the PDA has a lower dynamic range and higher noise. Its great value lies in its use as a simultaneous multichannel detector, having found its greatest application as an HPLC detector (Section 9.3.2).

DQ 18.3

Which factors limit the dynamic range of the photodiode array at:

(i) high light intensity;
(ii) low light intensity?

Answer

*At high light intensity, the capacitor will be **discharged quickly**, and the response of the PDA will cease when its voltage falls below a minimum level. At low light intensity, the response can be increased by extending the exposure time. However, this improvement will be limited by an increased dark-current response due to the **leakage of charge** from the capacitor.*

18.4 Charge-Transfer Devices

Why are charge-coupled detector cameras able to operate at such low light levels?

Recent developments in solid-state detection techniques have now produced very effective two-dimensional array detectors that operate on a charge-transfer process, as an alternative to photodiodes. These detectors have found widespread use in digital and video cameras, and their application to analytical instruments is now being commercially exploited.

The term 'charge-transfer device' (CTD) is a generic name that describes a detection system in which a photon, striking the IC semiconductor material, releases electrons from their bound state into a mobile state. The 'released' charges, consisting of negative electrons and positive holes (absence of electrons), then drift to, and accumulate at, surface *electrodes*. An array of these surface electrodes divides the detector into separate, light-sensitive, 'pixels'. The charge that accumulates at each electrode is proportional to the integrated light intensity falling on that particular pixel. The cycle of 'initialization, exposure and read-out' is similar to that of the photodiode array (Section 18.3.1).

An inherent problem with charge-transfer devices is that of 'blooming', which is observed as an apparent glow surrounding bright spots in the image. This can occur when excessive charge, created by bright light at one pixel, spills over into adjacent pixels, thus giving the glow. The effect of blooming can be reduced by creating physical barriers around areas in the chip where intense light is expected. It can also be reduced in charge-injection devices (CIDs) (see below) by draining excess charge away from those pixels that have been identified as receiving more light.

There are two distinct *classes* of device, (namely charge-coupled devices (CCDs) and charge-injection devices (CIDs)), which are differentiated by the *read-out process* used to record the charge packets that accumulate during the exposure time.

In a charge-coupled device, all of the charge packets are *moved* 'in-step' along the array row from one pixel to the next as in a 'bucket chain'. At the end of the row, the charge packets are fed sequentially into an *on-chip* low-noise amplifier, which then converts the charge into a voltage signal. The overall signal profile across the two-dimensional array is recorded one row at a time, thus giving a series of voltage signals corresponding to all of the pixels in the detection area.

Since all pixels in the CCD system are read in a single 'read-out' process, the exposure time for all of the pixels must also be the same. In addition, the read-out process is destructive — it removes the charge from the pixel.

In a charge-injection device, the charge accumulated in each pixel can be measured *independently* and *non-destructively* by using a network of 'sensing' electrodes which can monitor the presence of the accumulated charge. This is an important factor that differentiates CID systems from CCD and PDA systems in which the whole of the detection area is 'read' destructively in a single process.

The thermally generated dark current (see Section 18.1 above) in all of these devices can be reduced by cooling the detector. However, there is also electronic noise associated with the actual 'read-out' process.

The CCD system has very low 'read-out' noise, thus making it very suitable for very low light applications — look for example at the use of CCDs in video cameras that routinely operate with minimal illumination. The PDA and CID systems, which use the multiplex system of read-out, display considerably more noise.

The effective noise in the CID system can be reduced by *repeated* readings of the same accumulated charge and averaging out the different noise signals created in each read-out process. It is also possible to use 'binning', which is a process whereby the charge from adjacent pixels can be added together during the read-out process. The noise associated with *reading* the 'binned' charge is no greater that that of each 'un-binned' charge, although the aggregated signal will be greater.

As the CID system can measure the charges non-destructively, it is also possible to use an initial short exposure time to assess the spread of intensity across the different pixels. Those with low intensity are given a longer exposure time before measurement, while those with a very high intensity will be measured early, and any excess charge drained away to reduce 'blooming'.

Developments in this area of technology are extremely rapid. The latest CCD detectors have also been developed to allow different integration times for separate pixels.

DQ 18.4

Identify the feature(s) of a combined CID and echelle-grating system (Section 16.3) that make it particularly useful as a detection system for inductively coupled plasma-atomic emission spectrophotometry.

Answer

The echelle grating has a very high resolution (< 0.01 nm), and will resolve very close lines and also record the background signal in between the emission lines. Without any moving parts, the accuracy of the grating is excellent, and the full spectrum can be spread out in rows of pixels across the face of the CID detector, thus giving the ability to simultaneously measure at many different wavelengths.

By using different integration periods for different pixels in the CID detection process, it is possible to achieve a dynamic photometric range of several orders of magnitude. This enables the same measurement run to record both high- and low-intensity emissions across the spectrum. The CID system also has the ability to drain excess electrons to reduce blooming around the very intense lines.

18.5 Detectors of Infrared Radiation

What happened to the Golay cell?

IR detectors have been developed which use both thermal and photon detection processes. In the case of the thermal detectors, various methods have been used to detect the slight temperature increase (see Section 18.1 above), including the following:

- voltage generated by the temperature difference, (thermocouple);
- change of resistance (bolometer);
- expansion of gas (Golay detector cell);
- pyroelectric response.

Until recently, thermal detectors have displayed relatively slow response behaviour because of the time taken for the detector to reach an equilibrium temperature in reaction to a heat input, with typical time constants being measured in fractions of a second. However, new developments using small pyroelectric crystals, e.g. deuterated triglycine sulfate (DTGS), have dramatically reduced the response times. In addition, the improved speed of response of IR detectors helped to open up the practical possibilities of Fourier-transform infrared systems.

When the IR radiation falls on the pyroelectric crystal, a potential difference is generated with only very minimal current flow. Consequently, there is negligible shot noise, but, as with other thermal detectors, significant noise arises from thermal energy. The level of thermal noise can nevertheless be reduced by reducing the ambient temperature of the detector, using either liquid nitrogen or a Peltier-effect cooling device.

An important class of IR detectors, which use photon detection, are the MCT detectors. Mercury cadmium telluride (MCT) is a semiconductor material, whose *conductivity* increases due to the generation of mobile electrons by absorbed IR photons. The characteristics (responsivity and spectral range) of individual detectors depend upon the proportion of mercury and cadmium in the semiconductor matrix.

Although the MCT detector uses a 'photon' process, it is still susceptible to the creation of mobile electrons by thermal energy–Johnson (thermal) noise, and must be cooled to liquid-nitrogen temperature (77 K) to reduce this noise.

DQ 18.5

An FTIR spectrophotometer, when employed for basic measurements in the $4000-400$ cm^{-1} range, normally uses a DTGS detector, but when a microscope attachment is added, a MCT detector is required. Link the properties given in the table below to the appropriate detector:

Property	Detector	
	DTGS	MCT
Requires cooling to 77 K		
Greater dynamic range		
Less expensive		
Greater responsivity		

Answer

The MCT detector gives the greater responsivity required for the low light levels associated with microscope measurements, but it does require cryogenic cooling. For routine, non-microscope use, the cheaper DTGS detector, with its greater dynamic range, is often preferable.

SAQ 18.3

Identify which of the following are true advantages of an HPLC diode-array detector, and explain why:

(a) low noise due to the use of photodiodes   (True/False?);
(b) no moving parts   (True/False?);
(c) high wavelength accuracy   (True/False?);
(d) large dynamic range   (True/False?);
(e) rapid spectral acquisition   (True/False?).

SAQ 18.4

Give an example of a type of detector which is used routinely in analytical instruments to give three-dimensional data output.

Note. Further questions that relate to the topics covered in this chapter can be found in SAQs 11.5 and 14.2, and DQ 9.8.

Summary

This chapter has differentiated between the detection of radiation by using photon and thermal processes. It referred to the well-established types of vacuum-tube detectors, but also introduced modern developments in microelectronic array detectors, and compared the different types of new charge-transfer devices.

In addition, it looked at the performance characteristics of the detectors, particularly with respect to noise and dynamic range.

Responses to Self-Assessment Questions

Chapter 1

Response 1.1

Bias gives the same (possibly) unknown error every time the measurement is repeated. It is therefore necessary to use one of the following approaches:

- the same measurement procedure on a sample of known value (*reference material*) to estimate the bias;
- a method of known bias (*reference method*) on the same unknown sample, and then compare the results.

Imprecision at the level of repeatability can be assessed without external standards by repeating the measurement and observing the variation in results. However, imprecision at the level of reproducibility requires that the measurement is repeated by different laboratories, using different equipment and reagents.

Response 1.2

A concentration of 0.5 ppm is a fractional concentration of 0.5×10^{-6}. By using equation (1.1), we get an estimate for the coefficient of variance, $CV(\%)$, of $2^{(1-0.5(-6.3))} = 2^{4.15}$. This gives an estimated precision of 18%.

Response 1.3

In this present book, we define an analytical instrument as one in which an *experiment* is being performed (Section 1.2), the *conditions* of which are under the *control of the operator*.

Response 1.4

The expected answer is 'ruggedness'. This is a qualitative concept because the variation of the method may depend in a complex way on a variety of different factors (e.g. operator, reagent supplier, etc.).

An assessment of 'ruggedness' is important in predicting the reliability with which the same method can be used in different laboratories. However, the actual process of testing ruggedness can also help to development an understanding of the type of errors that may arise in the method.

Response 1.5

The *method* may use the instrument to compare an unknown sample with a known standard. In such a *comparative* method, it is the repeatability of instrument response between measurements (precision) that is most important for the accuracy of the overall analysis.

Chapter 2

Response 2.1

The signal values for random noise follow a Gaussian distribution. The rms value is equal to the standard deviation. If the standard deviation is σ, then 98.8% of values will be within $\pm 2.5\sigma$ of the mean value — this gives a range of 5σ. It is not possible to define absolute values for peak-to-peak values of a Gaussian distribution, but we see that only 1.2% of values will fall outside the 5σ range, and this then gives a reasonable working value.

Response 2.2

The figure is not given as a standard deviation. Hence, by taking ± 0.5 nm as limiting values, we can use equation (2.3) to evaluate the standard uncertainty, as follows:

$$u(\lambda) = 0.5/\sqrt{3} = 0.29 \text{ nm}$$

Response 2.3

The variation in pump flow rate does not have a first-order effect (see Section 9.4) on the measurement of peak area and can therefore be discounted in this calculation. However, the effect of variations in the photometric measurement and the autosampler contribute randomly to the uncertainty in the peak area. The relative photometric error is $0.008/0.8 = 1\%$, with the combined effect being calculated by using equation (2.7), as follows:

$$u(A) = \sqrt{(1.0^2 + 0.7^2)} \approx 1.2\%$$

Response 2.4

The combined uncertainties in the two cases are $\sqrt{(4^2 + 2^2)} = 4.5\%$ and $\sqrt{(4^2 + 1^2)} = 4.1\%$. The dominant factor on the overall uncertainty is clearly sampling, with the improvement in instrumental accuracy having only little effect. It would not be worth the time or cost to improve the instrumental measurement without first tackling the problem of sampling accuracy.

Response 2.5

We can express the uncertainty in T as $u(T) = \pm 0.005$. By using equation (2.13), we can derive equation (5.15). Substituting for $u(T)$, we obtain the following:

$$u(A) = \frac{0.005}{2.30 \times T}$$

The greatest uncertainty in the range $0A$ to $0.5A$ will occur for the largest value of A (i.e. the smallest value of T). When $A = 0.5$, $T = 0.32$, and hence we obtain:

$$u(A) = 0.005/(2.3 \times 0.32) = \pm 0.007A$$

for the range $0A$ to $0.5A$. Similarly, we can then obtain:

$$u(A) = 0.005/(2.3 \times 0.1) = \pm 0.022A$$

for the range $0.5A$ to $1.0A$.

Chapter 3

Response 3.1

The specifications of an instrument give a full description of all aspects of the instrument, including the performance characteristics, but they will also give other data e.g. dimensions, weight, power supply requirements, ease of maintenance, etc.

Response 3.2

It is interesting to note that the specifications are precise in specifying the exact *conditions* under which the measurements have been made, i.e. with water at a given flow rate and pressure. Pumping accuracy will vary with the fluid being used and the pressures involved.

However, the main point that should be noted is that there must be a *mistake* in the published values — accuracy (i.e. the estimation of *total* error) can not be less than precision (which is only one possible source of error).

Response 3.3

(i) The responsivity is quoted as coulombs per gram, and the minimum signal as grams per second, so the final signal will be coulombs per second, which is equivalent to amperes. The signal is therefore a current.

(ii) We can calculate the *change* in current due to the *LOD* signal as follows:

$$\Delta S_0 = 2.2 \times 10^{-2} \times 1.5 \times 10^{-12} = 3.3 \times 10^{-14} \text{ A}$$

However, the actual current will depend on what standing current, S_{blank}, is present as an offset current — see Figure 3.5.

(iii) If the *LOD* is given by $3\sigma_S/R'$, the *LOQ* is given by $10\sigma_S/R'$ (Section 3.3). The *LOQ* can then be estimated directly from the *LOD* as follows:

$$LOQ = 10 \times LOD/3 = 5.0 \times 10^{-12} \text{ g/s}$$

However, you should always treat such quoted values with caution.

Response 3.4

(i) The drift in the '0A' setting for a single-beam spectrophotometer can be compensated for by frequent recalibration (see Section 5.2) with a reference sample (blank), and avoidance of the measurement of very low and very high absorbance values. Note the large errors due to k_D (see Figure 5.7) at low and high values of A.

(ii) Imprecision in injection volumes in a GC autosampler can be compensated for by the use of a known amount of an internal standard added to the sample (spiking). The magnitude of the output signal for this standard can then be used to monitor the actual sample volumes being injected (see Section 10.2).

(iii) Poor output linearity can be compensated for by plotting a calibration graph with a range of known standards that bracket the value of the unknown.

(iv) Wavelength imprecision in a spectrophotometer can be compensated for by making absorbance measurements at the peaks of the absorbance lines (see Section 15.5.1).

The complexity of the analytical method can be reduced if the performance reliability of the instrument can be improved. Nevertheless, it is essential that the analyst is aware of the performance capabilities and limitations of the instrument, and takes these fully into account in the design of the analytical method.

Response 3.5

Many instruments (e.g. spectrophotometers) perform an *experiment* that has applicability in very many different analytical procedures, and where the detection limit is dependent on the procedure itself rather than principally on the instrument.

However, for some instruments, it is possible to quote useful detection limits (Section 3.3), provided that:

- the detection limit applies to a specific analyte;
- the instrument itself sets the major limit on detectability.

Good examples of this include the following:

(a) *Atomic spectroscopy*, which measures atomic concentration, with the detection limits being a function of the element rather than the sample matrix (see Section 6.5). Note, however, that for GF-AAS, the detection limit is too dependent on variations in the particular sample matrix for it to be quoted for general applicability.

(b) *Chromatography*, where many detectors (see Section 9.3 and 10.3) have a specific selectivity for a well-defined range of analytes. It is thus possible to define, for the specific detectors, the detection limits for the specific analytes.

Chapter 4

Response 4.1

It is unwise to plan *unnecessary* testing as this costs time and money, and it also undermines the credibility of the quality assurance programme. Clearly, it is necessary to confirm performance after any maintenance work has been done on the instrument, and this contributes to the quality assurance of performance into the future. However, testing *before* maintenance can also be of real benefit, either as a means of confirming the quality of performance in the period leading up to maintenance, or as a 'before and after' check on the impact of the maintenance itself. The need for the 'before' check depends on the way in which the quality assurance programme is designed.

Response 4.2

It is an essential part of any quality system that the item or service should be 'Fit for Purpose' in respect of its 'Future Intended Use'. It is essential to define carefully the 'intended use' before its is possible to confirm that the service is actually fit for that use. Too often, laboratories carry out tests which do not address the particular *decision* that the client needs to make, and provide data that are not necessarily appropriate, or have an inappropriate level of accuracy, for the given problem. The VAM principle ensures that both the analyst and the client agree on what is to be done.

Response 4.3

System Suitability Testing (SST) is specifically a 'top-down' process of validation. It occurs as part of a validated analytical method, and confirms, at the time of use, that the experimental apparatus has a performance adequate for the analysis being carried out. Equipment Qualification (EQ), however, uses a mixture of 'top-down' validation processes (Performance Qualification' (PQ) and Operational Qualification (OQ)) and 'bottom-up' qualification processes (Design Qualification (DQ) and Installation Qualification (IQ)) to provide an overall 'bottom-up' confirmation of the 'fitness for purpose' of the instrument or module.

Response 4.4

The answer will depend on several factors that you should have addressed *earlier* as part of a quality assurance programme. Examples include the following:

- Does the performance fall outside limits that you have determined as necessary for your own analysis, i.e. your user requirement specification (URS)? If the instrument continues to satisfy these specifications, then it is still OK for use.

- Has the change in performance occurred suddenly? A knowledge of this would have required an ongoing regime of performance monitoring. If the change is sudden, then it would be wise to look for the possible cause. Otherwise, it can be treated under routine maintenance.

Response 4.5

The noise level in %T may have a direct effect on an analytical measurement and may form part of both 'top-down' checks (e.g. system suitability testing) and 'bottom-up' checks (e.g. equipment qualification and quality control checks as part of a maintenance programme). The energy at the detector is an important 'bottom-up' indicator of instrument 'health' and will be tested as for noise in equipment qualification, but it may not feature in 'top-down' tests such as system suitability testing.

Chapter 5

Response 5.1

The first step is to calculate (using equation (5.3)) the apparent transmittance, $T_{app} = 10^{-A} = 0.0063$. Assuming that the stray light actually has the maximum quoted figure of $0.05\% = 0.0005$, we can calculate the true transmittance, i.e. $T_{true} = T_{app} - S_L = 0.0063 - 0.0005 = 0.0058$. The true absorbance can then be calculated, i.e. $A_{true} = -\log(T_{true}) = 2.236$.

We can then say that the maximum error due to stray light at this absorbance is − 0.036 or − 1.6%. Notice that the error due to stray light always gives a reduced absorbance.

The percentage error due to stray light increases with the absorbance of the sample.

Response 5.2

By referring to Figure 5.7, at the level of fractional uncertainty (= 0.004), it can be seen that acceptable uncertainty is obtained between absorbance values of about 0.09A and 1.75A. This gives a dynamic range of 1.75/0.09 = 19. However, this is a rather contrived example which is presented to illustrate the point and some of the values are not necessarily typical.

Response 5.3

The relative noise in the absorbance signal will decrease with an increase in signal energy due to a wider spectral bandwidth, while it will increase with decreased signal energy due to an absorbing sample. Hence, a possible allocation of the values is as follows: (a) < 0.0004A; (b) < 0.0002A; (c) < 0.002A; (d) < 0.0008A.

Response 5.4

The problem with scanning in a spectrophotometer is the need to compare test and reference samples at every wavelength, coupled with the fact that the source intensity can drift between measurements. Early scanning spectrophotometers required a double-beam system in order to obtain a continual comparison between the test and reference samples. With the development of more stable sources and the use of computer memories for spectral data, it is now possible to use scanning single-beam systems with consecutive measurements of reference and test spectra. Modern diode-array systems now also permit rapid parallel acquisition of the spectra, and commercial Fourier-transform UV−visible systems are also being developed.

Response 5.5

We can see from Figure 5.7 that the relative uncertainty in the measurement of absorbance (and hence concentration) depends on the conditions under which the measurement is made, i.e. it would not be appropriate to attempt to measure the concentration of a sample with a very high or very low absorbance. The analytical method should aim to produce samples with absorbances near to the minima in the graph.

Chapter 6

Response 6.1

The system needs to be optimized (e.g. selection of lamp, flame height, temperature, etc.) for each element to provide the best results, and for most systems it is more convenient to set the optimization for one element and then work through all of the samples before proceeding to the next element. However, some instruments provide rapid switching and optimization between different element lamps, so that it can become more efficient to measure several elements in the same sample before proceeding to the next sample.

Response 6.2

The flame system suffers from poor nebulizer efficiency, with only a small fraction of analyte reaching the flame, coupled with a large dilution of the analyte in the flame itself. GF-AAS ensures that all of the analyte is atomized and is presented into the radiation beam.

Response 6.3

'Linearity' (Section 3.2.4) refers to the photometric range over which the responsivity does not deviate by more than a given factor, which is unspecified here but would normally be 5%. Note that deviations at very low absorbance values are not normally included (Section 5.8.3) for a photometric response range.

'Precision' has been measured by taking the standard deviation as a percentage of the value (relative standard deviation) for 10 integrated (see Section 14.5.5) readings.

Response 6.4

In ICP-AES, each atom may emit radiation at several different *wavelengths*, corresponding to different energy transitions within the atom. In ICP-MS, the spectrum is in units of m/z, and each atom may produce multiple signals due to the presence of different isotopes and/or multiply charged atoms. Other interferences include background radiation for ICP-AES, and polyatomic atoms (with argon and oxygen) originating from the plasma in ICP-MS.

Chapter 7

Response 7.1

The minimum length of travel *in opd* for a resolution of 0.5 cm^{-1} is given (Section 7.4.2) by $1/0.5 = 2$ cm. At an opd speed of 0.5 cm/s, this would take 4 s. In practice, additional data may be collected to allow the software to compensate for phase shifts caused by the beam-splitter, and the instrument will also need to perform a fast Fourier transform.

Response 7.2

All measurements are made by using the same spectral bandwidth (4 cm^{-1}), with either 4 s or 1 min being allowed to obtain the data. The effect of taking 1 min instead of 4 s allows for 15 times more repeated scans to average the noise. Thus, it could be expected that the noise figure for 1 min integration time will be $\sqrt{15}$ times better, which is an improvement by a factor of 3.87.

Instrument (c) uses the root-mean-square (rms) value for noise, which will be less than the peak-to-peak value, thus giving an apparently enhanced ratio between signal and noise. We can use a factor of about '5' for the ratio between the peak-to-peak and rms values for a random noise signal (see SAQ 2.1).

If we apply both corrections, it gives very similar S/N ratios for all of the instruments.

Response 7.3

For the UV–visible system, a decrease in spectral bandwidth is achieved by reducing the slitwidth of the monochromator, and this then reduces the signal power as the square of the bandwidth (see Section 16.2.4). As a direct consequence, the S/N ratio will also decrease.

For the FTIR system, a decrease in spectral bandwidth is achieved by increasing the length of the mirror travel (Section 7.4.2), which will increase the time taken for the recording of each interferogram. As a result, provided that the overall time limit for measurement is not changed, there will be fewer repeat measurements available to average the noise (Section 7.5), and the S/N ratio will therefore decrease.

Response 7.4

Due to the low energy of IR photons, the detectors are inherently more susceptible to thermal noise (see Section 14.4.1). Consequently, it is important to ensure, by measuring energy at the detector, that the interferometer is adjusted for maximum energy transmittance. The noise level itself is also monitored.

Response 7.5

Without correction, the instrument produces a 'sinc' instrument lineshape (7.4.3). This lineshape has the disadvantage of having pronounced side lobes and gives some apparently negative transmittance values. The apodization control will apply a software convolution correction to the interferogram before it is transformed into the IR spectrum, which will produce a more acceptable lineshape for narrow spectral lines. The disadvantage of apodization is that it will also broaden very narrow lines, and a compromise must therefore be reached.

Chapter 8

Response 8.1

A **liquid** (see Section 9.2) has much lower compressibility and is less prone to leakage than a gas. Thus, the volume flow rate of a liquid can be adequately controlled by a piston-pumping system that sweeps out well-defined volumes. Sophisticated pumping systems can also provide compensation for the different compressibilities of various solvents. A problem in continuously pumping liquids is that of pulsations due to reciprocation in the pumping system, but this can be reduced by pulse dampers (which increase dead volume) and the use of multiple pistons.

The flow rate of a **gas** (see Section 10.1.1) is controlled by varying the applied pressure, while monitoring the observed flow rate. However, the characteristics of a gas are very dependent on both pressure and temperature, which makes it difficult to measure the flow rate accurately. Manual systems normally use a constant-pressure setting and a bubble flowmeter attached at the end of the system, while modern electronic systems now provide a more accurate measurement of gas flow, and give different options for electronic flow control.

The flow rate in **capillary electrophoresis** (see Section 11.1) is driven by electro-osmotic flow, and is not dependent on any pressure applied externally to the column. The rate here is determined by the applied voltage, but is also sensitive to the properties of the electrolytic buffer.

Response 8.2

The *most likely* distance covered under random-walk conditions increases as the square root of the number of steps (see Section 13.5). By comparison, the spread of the component will increase as the square root of the time available for the diffusion. The width of the peaks will therefore increase in proportion to the square root of the retention time (a) (see also Section 8.4).

Response 8.3

The unretained void time, t_M, is 1.15 min, with the data for peaks P and Q being as shown in the following table:

Parameter	Peak	
	P	Q
t	4.0	4.95
W_H	0.35	0.35
$k = (t - t_M)/t_M$	2.48	3.30

The separation factor, $\alpha = k_Q/k_P = 1.33$, while the resolution, $R_S = 1.18(4.95 - 4.0)/(0.35 + 0.35) = 1.60$.

By substituting these values into equation (8.15), it is possible to calculate N as follows:

$$1.60 = (\sqrt{N}) \times 0.25 \times (3.3/4.3) \times (0.33/1.33)$$

thus giving $N = (33.6)^2 = 1130$. This equates very approximately to the value (1110) calculated in DQ 8.2. The uncertainties in this type of calculation can be explained by the approximate values used for the peak widths, coupled with the fact that N is finally obtained by squaring the calculated values.

Response 8.4

The area of the peak is often proportional to the amount of analyte, and provided that the peak width remains the same, the peak height is also proportional to the peak area. However, it is generally preferable to measure the peak *area* because data are being used from across the peak, rather than concentrating on the data points just around the peak.

It is particularly important to measure area in chromatography because of possible variations in peak width and shape which would affect the accuracy of the peak-height method. The shapes of spectrophotometric lines are more reproducible (being dependent on the analyte rather than the process of measurement), and hence peak height is a more reliable measure than in chromatography. When measuring the area of a spectral line, care has to be taken that the summation of the line is carried out in absorbance and not transmittance — see Section 15.5.

Chapter 9

Response 9.1

(i) If we calculate the accuracy on the basis of $\pm 0.05\%$ of the maximum flow rate, we obtain a value of $\pm 0.05 \times 50/100 = \pm 0.025$ ml/min. We can then compare this with the $\pm 1.00\%$ values of the selected flow rates for the various settings, as shown in the following table:

Flow-rate parameter	Accuracy at various flow-rate settings (ml/min)			
Flow-rate setting	0.05	0.5	5	50
$\pm 0.05\%$ of maximum flow (50 ml/min)	± 0.025	± 0.025	± 0.025	± 0.025
$\pm 1.00\%$ of flow setting	± 0.0005	± 0.005	± 0.05	± 0.5
Larger value	± 0.025	± 0.025	± 0.05	± 0.5
Percentage accuracy	$\pm 50\%$	$\pm 5\%$	$\pm 1\%$	$\pm 1\%$

(ii) Clearly, the greatest relative inaccuracy arises at the low flow rates. Note, however, that the *repeatability* of the flow rate is often a more important parameter than absolute accuracy, and this repeatability is likely to be considerably better than the quoted accuracy.

(iii) Although each pump has a very wide dynamic flow range, the dynamic range *at the maximum relative accuracy* is not so wide, and the effective overlap is therefore not so great.

Response 9.2

It is expected that the pressure may change with the *changing viscosity* of the mobile phase, as the proportions of different solvents vary with the chosen gradient. When the run is repeated, the pressure should follow the same cyclical variation.

Response 9.3

A negative peak is clearly due to a detection phenomenon. Two possibilities are that the analyte has a lower absorbance than the mobile phase in a UV detector, or that it has a lower refractive index in a refractive-index detector.

Response 9.4

The effect of the wavelength of the UV light is clearly a key factor and we see that the problem must also be related to the solvents used (because of the effect of the gradient). We know that some solvents can absorb at short UV wavelengths (Section 9.3.1), and an absorbance in the solvent will thus give an apparent change in the baseline. If one of the solvents in the gradient mixture (Section 9.2.4) is absorbing more than the other, then the absorbance of the mixture will depend on the composition of the mixture itself. A changing mixture (gradient elution) will give a changing baseline.

If, however, (i) *there is no change of composition* (isocratic), or (ii) *the measurement is at wavelengths where there is no absorbance*, then the absorbance will not change and the baseline will stay at the same value. This is consistent with the observed symptoms.

Response 9.5

One possibility is that the operator may have increased the time constant on the detector filter.

Chapter 10

Response 10.1

The first method is clearly temperature programming of the column. However, the high final temperatures will tend to increase column bleed. With modern electronic control of gas flows, it is also possible to *increase* the gas flow towards the end of the run in order to reduce the later retention times. Note that, at constant pressure, the increased temperature actually reduces the gas flow rate.

Response 10.2

The FID signal, *integrated in time*, is proportional to the total mass of analyte as it passes through the flame, and is therefore independent of flow. The signal from the TCD is proportional to the concentration of the analyte in the gas and is independent of the flow rate of the gas. If the gas moves slowly, then the same signal will be integrated for a longer time, thus giving a larger integrated signal.

Response 10.3

Signal filtration refers to the use of a low-pass analogue filter which is used to smooth noise. The figures will apply to the time constant (see Section 13.6.2) of the filter, and a reasonable guess would suggest that the units are in *milliseconds* (a time constant of 800 s would be unreasonable!).

Response 10.4

(i) Sensitivity in this case is equivalent to *responsivity*, i.e. the change of output signal (μV) as a function of the change of input signal (concentration of nonane in ppm). However, the minimum detectable quantity (*MDQ*) is an indication of the *minimum* concentration of nonane that can be measured.

(ii) For sensitivity, the *conditions* of the measurement are not given (e.g. temperature of the filament), while for the *MDQ*, the *ratio* between the minimum detectable signal and the standard deviation of repeatability is not stated. We could assume that the latter value is '3', but some manufacturers use different values, e.g. S/N = 2. In addition, there is no indication of the acceptable deviation from *linearity* that is used to define the linearity range of 10^5 — it may be 5%, although it is not specified.

Chapter 11

Response 11.1

The electric field strength is $(25 \times 10^3/35)$ V/cm $= 0.714 \times 10^3$ V/cm. The EOF velocity is $25/(2 \times 60)$ cm/s $= 0.208$ cm/s; similarly, the velocity of the analyte is 0.0926 cm/s.

The observed mobilities are then $0.208/(0.714 \times 10^3) = 2.91 \times 10^{-4}$ cm^2/V s, and $0.0926/(0.714 \times 10^3) = 1.30 \times 10^{-4}$ cm^2/V s.

Thus, the mobility of the analyte is $(1.30 - 2.91) \times 10^{-4}$ cm^2/V s $= -1.61 \times 10^{-4}$ cm^2/V s.

Response 11.2

The effective mobilities, μ_{eff}, in the two cases are (i) $5 \times 10^{-4} \text{cm}^2/\text{V s}$, and (ii) $1 \times 10^{-4} \text{cm}^2/\text{V s}$, respectively.

The change in effective mobility, in both cases, due to the change in μ_{EOF}, is $0.02 \times 3 \times 10^{-4} \text{ cm}^2/\text{V s} = 0.06 \times 10^{-4} \text{ cm}^2/\text{V s}$. This gives fractional errors of (i) $100 \times (0.06/5) = 1.2\%$, and (ii) $100 \times (0.06/1) = 6\%$.

Note the large error when the mobility of the negatively charged ions approaches the magnitude of the EOF mobility.

Response 11.3

If the width at half-height is $0.15 \text{ min} = 9 \text{ s}$, then the standard deviation of the peak is $9/(2.355) = 3.82 \text{ s}$ (see equation (8.5)).

An injection time of 4 s can be equated (DQ 11.4 and equation (2.3)) to a distribution with a signal standard deviation, σ_S, of $4/\sqrt{12} = 1.15 \text{ s}$.

The standard deviation, σ_R, due to the *other* causes of line broadening, can be estimated from substituting into equation (11.7), as follows:

$$\sigma_{\text{tot}} = \sqrt{\left(\sigma_R^2 + \sigma_S^2\right)}, \text{ i.e. } 3.82 = \sqrt{\left(\sigma_R^2 + 1.15^2\right)}$$

which gives $\sigma_R = 3.64$. Thus, injection broadening is not a significant cause of the final peak width.

If σ_S is doubled to 8 s, the new peak width can be recalculated as follows:

$$\sigma_{\text{tot}} = \sqrt{(3.64^2 + 2.3^2)} = 4.3 \text{ s}$$

which is a slight increase.

Response 11.4

The area of a spectrophotometric absorbance line is proportional to quantity (Beer's Law), but until the development of diode-array detection (see Sections 18.3.1 and 9.3.2) it had not been convenient to produce an integrated value for the line area. This is due in part to the fact that a traditional spectrophotometer actually measures transmittance (Section 5.1), and it is then necessary to use a narrow spectral bandwidth in order to preserve photometric linearity (see Section 15.5.1). The measurement is therefore normally made at the wavelength of peak absorbance. However, the use of diode-array detection does now allow bunching of absorbance values (Section 9.3.2).

It is convenient to integrate the signal from a chromatographic detector as the signal arrives, and this area will be proportional either to quantity or concentration (Section 8.6). In capillary electrophoresis, the time that the component spends in an on-column detector will be proportional to its migration time, and hence this will affect the result from a detection process that measures concentration (Section 11.3). A correction is made by dividing the peak area by the migration time.

Response 11.5

They all use a 'reverse' process of detection. A standing signal is established which is **reduced** in the presence of the analyte. The problems with this type of detection are that (i) the very low level analyte signals are subject to the noise created by the large standing signal, and (ii) large analyte signals are in danger of saturating the detector by reducing the standing signal to near zero. The two effects limit the potential dynamic range of such detectors.

Chapter 12

Response 12.1

It will be necessary to ensure that the ions are produced with multiple charges, i.e. with a z-value of at least 15. This can be achieved by using electrospray techniques.

Response 12.2

Yes *and* No! Resolving power in all cases gives the ratio of the peak (or line) position to the peak (or line) width, and, as such, indicates the precision with which the signal has been defined *in position*. The term indicates the *selectivity* of the process of separating wavelength or mass components, or components with different retention times. The difference in the use of the terms is in the definition of peak width. For chromatography, the peak width is given as the standard deviation (Section 8.8), while for the monochromator it is usually the spectral bandwidth (see Section 16.7). However, the '10% criterion' for mass spectroscopy requires that the peaks should be separated by five standard deviations.

Response 12.3

ICP-MS definitely enables measurements to be made at very low concentrations (a, true). Although this technique has a very wide dynamic range, the lower limit is so low that the upper limit still requires dilution of more concentrated samples (b, false). The attraction of the wide *photometric* dynamic range must be treated carefully, because there are other interference factors that can hinder the simultaneous, or sequential, measurement of high and low concentrations, such as spectral interferences, memory effects, etc.

Response 12.4

It is possible to assess the effect of interferences from other elements, and by using the knowledge of relative abundances, to make quantitative corrections.

Chapter 13

Response 13.1

The process of converting the digital number to its analogue equivalent is a simple matter of multiplying each 'bit' by its binary weighting and then adding the values together. However, try the reverse process of converting, e.g. 679, into a digital number, and then explain how you did it to a friend or colleague. It is not so easy mathematically, and it is also not so easy electronically. This explains why analogue-to-digital converters are generally much slower than digital-to-analogue converters.

Response 13.2

The two waves will keep going 'in and out of phase', alternately producing constructive and destructive interference. The waveform will be similar to that shown in Figure 7.7. The result will sound to the ear as though the intensity of a 1001 Hz signal is fluctuating with a frequency of 2 Hz. This is the phenomenon known as *beats*, which can be used to detect if musical instruments are out of tune.

Response 13.3

Your Fourier transforms for P and Q should look the same as that shown in Figure 13.5. The width of peak Q is 0.235 min = 21 s, which will give a frequency bandwidth of about 0.02 Hz (equation (13.6)). The width of peak P is slightly greater, which would then give a slightly narrower bandwidth.

The main difference between the Fourier transforms of P and Q is that their relative displacement in *time* would be represented by a change of '*phase*' of the components. However, this phase difference is not likely to appear on your diagrams.

Response 13.4

The signal is non-periodic, so we can expect a spread of frequency components. We can then use equation (13.6) to estimate the width of this spread, Δf, and for a very short time, Δt, we see that the frequency components will be spread over a very wide frequency range. White noise, which has a similar width of spread of frequency components, is actually due to impulse events, e.g. when electrons move randomly due to thermal noise and shot noise.

Response 13.5

By using equation (13.6), we can see that the frequency bandwidth at half-height of the Fourier transform is $0.44/1 = 0.44$ Hz. Figure 13.5(b) shows that significant frequency components extend to frequencies well beyond this value. The question tells us that we should allow for frequency components up to three standard deviations, i.e. $3.5 \times \Delta f$ (by using equation (8.5)) $= 1.6$ Hz. The sampling frequency must be at least twice the frequency of the highest frequency component (Nyquist theory), therefore giving 3.2 Hz.

This is the minimum frequency which should be used to avoid aliasing, but if we wished to use the data directly to measure peak height and area, we would probably sample at a frequency of at least three times this figure.

Chapter 14

Response 14.1

It all depends on whether the signals are random or not. It may be that the bubble is oscillating and producing a regular pattern — this would be 'interference'. However, random spikes would be 'noise'. In either case, the solution is not to try and extract the spurious signals from the chromatogram, but to remove the bubble from the detector and ensure that no more are formed.

Response 14.2

(i) Ohm's Law tells us that the potential difference across the resistor, V, is given by $V = I \times R$. Therefore, we expect to measure a constant *wanted* analytical signal, V, of 0.5 mV.

(ii) We would like the constant analytical signal to be the only signal that we would record in the system. However, there will be additional *unwanted* signals, arising for a variety of different physical reasons, as follows:

- the electrons in the resistor have thermal energy which leads to a fluctuating thermal noise voltage (Section 14.4.1), $V_{N,rms}$, of 0.029 mV;

- the current is carried as discrete charges, thus giving a flow of charge that also has small random fluctuations due to shot noise (Section 14.4.2), $i_{N,rms}$, of 0.0057 pA;

- the value of the resistance itself will be changing due to a variety of factors (e.g. loss of moisture as the resistor warms up) which are more pronounced at lower frequencies — drift (Section 14.3);

- the resistor itself may be acting as a small aerial and picking up electromagnetic signals from the environment, which then appear as small voltage signals — interference (Section 14.5.2).

Response 14.3

We need to check that the time constant of the filter is not more than one tenth of the width of the peak, i.e. 0.1 min = 6 s.

The time-constant of the filter is calculated from the bandwidth by using equation (13.13), which gives values of (a) 0.016, (b) 0.16 and (c) 1.6 s, respectively.

Filters (a) and (b) are satisfactory, but the use of (c) would lead to signal distortion and it would be irrelevant to calculate a noise-reduction figure in this case.

For (a), the bandwidth is decreased by a factor of 10. Hence, the noise decreases by a factor of $\sqrt{10}$ and the S/N ratio increases by the same proportion to 31.6. For (b), the bandwidth is decreased by a factor of 100 and the time constant of the peak is still less than one tenth of the peak width — the S/N ratio will therefore increase to $10 \times \sqrt{(100)} = 100$.

Response 14.4

The process of 'averaging' the signals (Section 14.5.5) will improve the S/N ratio in the aggregated signal (a, true). However, as 'n' increases, there are fewer *aggregated* data points per second available to record the chromatogram. Therefore, narrow peaks will reduce the maximum value for 'n' that is consistent with keeping a sufficient number of points to define the peak (b, true). The speed of recording the chromatogram is a function only of the retention time of the peaks (c, false). The signal from every photodiode in the array will be recorded at 100 times per second — each diode signal is effectively recorded in parallel and independent of each other (d, false).

Chapter 15

Response 15.1

The Fourier transform of a rectangular function (as for the transmittance of a slit) is a sinc (x) function, while the Fourier transform of a triangular function is a sinc^2 (x) function. The effect of convoluting two rectangular functions (see Figure 16.5) to produce a triangular function is the same as multiplying their Fourier transforms (Section 15.3).

Response 15.2

Both processes are used to improve signal-to-noise ratios. Bunching can be used when the original number of data points collected is greater than the number required to specify (cf. the Nyquist criterion) the shape of the analytical signal, e.g. data collection from a diode array. When it is necessary to keep the original density of data points, the 'moving average' of the box-car convolution filter can be used.

Response 15.3

You must use the curve for $\Delta_I/\Delta_S = 8/20 = 0.4$ from Figure 15.5(a), in order to obtain the reduction from the true absorbance, A_P, as a percentage ($x\%$) of the observed absorbance, i.e. $A_O = A_P(100/(100 + x))$. The results obtained are as follows:

A_P	0	1	2	3	4
A_O	0	0.93	1.83	2.72	3.59

The plot produced by using these data shows a *negative* curvature.

Chapter 16

Response 16.1

The linear dispersion is $\Delta x/\Delta\lambda$. Hence, $\Delta\lambda = (0.5 \times 10^{-3})/(0.5 \times 10^6) = 1.0 \times 10^{-9} = 1.0$ nm.

The resolving power is $\lambda/\Delta\lambda = 500/1 = 500$.

Response 16.2

The total energy transmitted is proportional to the square of the bandwidth. Hence, this will drop by a factor of $2^2 = 4$.

The peak transmittance is not affected by the change in bandwidth and should remain at 80%.

Response 16.3

No monochromator is perfect. In practice, there will always be some transmitted energy through the monochromator at wavelengths outside of the designed spectral bandwidth (stray light). However, stray light can be reduced considerably if the light is already nearly monochromatic before it enters the monochromator. This can be achieved if two monochromators are connected in series — a double monochromator — with the output from the first providing the near monochromatic input for the second. The spectral bandwidth is reduced slightly, but the stray light is very significantly reduced. Spectrophotometers using a double-monochromator system would usually have a very wide photometric range.

Response 16.4

(i) Each spectral 'strip' represents an adjacent 'order' from the grating. In Figure 16.7, the prism gives *horizontal* dispersion. Point B is in the same *vertical* line as point A, and will therefore have the same wavelength i.e. 324 nm.

For a given angle of *vertical* dispersion from the grating, the product of order × wavelength is constant (equation (16.3)). Point C will be in the 51st order, but since it has the same '$n\lambda$' product as point A, we know that $50 \times 324 = 51 \times \lambda_C$, and therefore $\lambda_C = 318$ nm.

(ii) From equation (16.5), we can see that the linear dispersion increases with the order number, n. From Figure 16.7, it can also be seen that the higher spectral orders are used to produce the shortest wavelength. Thus, it is the shortest wavelengths (UV) that have the greatest dispersion.

Chapter 17

Response 17.1

The 'switching-on' process for a lamp is the most stressful time. Initially, the filament is cold and has a very low resistance. This results in a very large initial current that heats the filament quickly to its normal operating temperature, in which state the resistance increases and the current falls to a lower operating level. If the filament has already become narrow at one point due to evaporation of tungsten, the high initial current creates excessive local heating, thus causing further rapid evaporation and a possible 'explosive' break.

The lifetime of a lamp can be extended by reducing the number of times that it is switched on and off.

Response 17.2

The source types will be similar to those used in a 'stand-alone' spectrophotometer. Deuterium is a possible source for the UV range, while tungsten is recommended if the visible range is also being covered. Alternatively, a xenon lamp now provides a source that can cover both of these ranges.

Response 17.3

The spectral bandwidth and wavelength accuracy of the instrument depends on the monochromator and is not affected by the deterioration of the lamp (a and f, false). The lamp itself will not generate high-frequency noise (c, false). However, the fall in signal energy will reduce the S/N ratio in the instrument and the output of the *instrument* may then appear noisier (e, true).

A tungsten lamp gives a strong output at 650 nm, and it is unlikely that some loss of energy at this wavelength will seriously affect the photometric performance (b, false). The major effect is likely to be at short wavelengths (e.g. 350 nm) where the output from the tungsten lamp is already very low (Section 17.2). With a serious loss of energy, it may not even be possible for the output from the lamp to provide enough energy to reach the 100%T (or $0A$) setting for the reference sample (d, true).

Chapter 18

Response 18.1

The reason for cooling a detector is to reduce thermal noise (Section 14.4.3). This is appropriate when the quantum energy, $h\nu$, of the radiation photons is not significantly larger (see DQ 18.1) than the thermal energy, kT. In analytical instrumentation, this is particularly significant for IR wavelengths.

Shot noise in detectors, however, is not affected by cooling (Section 14.4.3).

Response 18.2

The total power in the visible range entering the monochromator is $20 \times 0.1 \times 0.15 = 0.3$ W. The wavelength of the monochromator is set to the 'mid-range' in the visible spectrum, so we can estimate that the power at the wavelength is approximately equal to the average over the spectrum (see Figure 17.1). Thus, with a 50% efficient monochromator and a bandwidth of 2 nm, we can estimate that the power leaving the monochromator will be $(0.3 \times 0.5 \times 2)/(650 - 350) = 0.001$ W. The sample with an absorbance of $2A$ will transmit 1% of the light (see equation (5.3)), thus giving a total power at the detector of 10^{-5} W.

This quantum energy of a 500 nm photon is given by hc/λ (see equation (17.1)) $= 3.96 \times 10^{-19}$ J. Thus, there are $10^{-5}/(3.96 \times 10^{-19}) = 2.53 \times 10^{13}$ photons arriving at the detector per second, which gives (allowing for the quantum efficiency of the detector) $0.08 \times 2.53 \times 10^{13} = 2.02 \times 10^{12}$ electrons per second. This gives a final current of $2.02 \times 10^{12} \times 1.6 \times 10^{-19}A \approx 0.3$ µA.

Response 18.3

The use of a diode-array detector does not confer any **photometric** advantages (Section 18.3.1) over other UV–visible detectors, and therefore (a) and (d) are both false. A key advantage is the speed (e) with which the full absorbance spectra of the various analyte components can be recorded. This enables a three-dimensional plot of absorbance against both wavelength and retention time to be produced. The other main advantage is the very good wavelength reproducibility (c), which permits a sophisticated analysis of overlapping spectra. The fact that there are no moving parts (b) contributes to the good wavelength reproducibility (Section 16.2.1) and to the ease of maintenance.

Response 18.4

One obvious example is that of the diode-array detector (Section 18.3.1) which can plot absorbance against both wavelength and time for an HPLC chromatogram (Section 9.3.2). An alternative answer would be the mass-selective detector (MSD), which plots quantity against both mass and time in GC–MS, LC–MS and CE–MS.

Bibliography

General Instrumentation

Ewing, G. W. (Ed.), *Analytical Instrumentation Handbook*, 2nd Edn, Marcel Dekker, New York, 1997.

Pungor, E., *A Practical Guide to Instrumental Analysis*, CRC Press, Boca Raton, FL, USA, 1994.

Robinson, J. W., *Undergraduate Instrumental Analysis*, 5th Edn, Marcel Dekker, New York, 1994.

Rubinson, K. and Rubinson, J. F., *Contemporary Instrumental Analysis*, Prentice Hall, Englewood Cliffs, NJ, USA, 1999.

Skoog, D. A., Holler, F. J. and Nieman, T. A., *Principles of Instrumental Analysis*, 5th Edn, Saunders College Publishing, New York, 1997.

Strobel, H. A. and Heineman, W. R. (Eds), *Chemical Instrumentation*, 3rd Edn, John Wiley & Sons, New York, 1989.

Willard, H. H., Merritt, L. L., Dean, J. and Settoe, F. A., *Instrumental Methods of Analysis*, 7th Edn, Wadsworth, Belmont, CA, USA, 1988.

Specific Techniques

Baker, D. R., *Capillary Electrophoresis*, John Wiley & Sons, New York, 1995.

Barker, J., *Mass Spectrometry*, 2nd Edn, ACOL Series, John Wiley & Sons, Chichester, UK, 1999.

Fowlis, I. A., *Gas Chromatography*, 2nd Edn, ACOL Series, John Wiley & Sons, Chichester, UK, 1995.

Khaledi, M. G. (Ed.), *High Performance Capillary Electrophoresis*, John Wiley & Sons, New York, 1998.

Kitson, F. G., Larsen, B. S. and McEwen, C. N., *Gas Chromatography and Mass Spectrometry*, Academic Press, London, 1996.

Lindsay, S., *High Performance Liquid Chromatography*, 2nd Edn, ACOL Series, John Wiley & Sons, Chichester, UK, 1992.

Meyer, V. R., *Practical High Performance Liquid Chromatography*, John Wiley & Sons, Chichester, UK, 1994.

Quality

CITAC, *International Guide to Quality in Analytical Chemistry — An Aid to Accreditation*, ISBN 0-948926-09-0, CITAC CGI, 1995. [Copies available from VAM Helpdesk, LGC (Teddington) Ltd, Teddington, UK.]

Ettre, L. S. and Hinshaw, J. V., *Basic Relationships of Gas Chromatography*, Advanstar, Cleveland, OH, USA, 1993.

EURACHEM, *Quantifying Uncertainty in Analytical Measurement*, ISBN 0-948926-08-2, 1995. [Copies available from VAM Helpdesk, LGC (Teddington) Ltd, Teddington, UK.]

EURACHEM, *Quality Assurance for Research and Development for Non-Routine Analysis*, ISBN 0-948926-11-2, 1998. [Copies available from VAM Helpdesk, LGC (Teddington) Ltd, Teddington, UK.]

EURACHEM, *The Fitness for Purpose of Analytical Methods*, ISBN 0-948926-12-0, 1998. [Copies available from VAM Helpdesk, LGC (Teddington) Ltd, Teddington, UK.]

Prichard, E. (Co-ordinating Author), *Quality in the Analytical Chemistry Laboratory*, ACOL Series, John Wiley & Sons, Chichester, UK, 1995.

Wilson, S. and Weir, G., *Food and Drink Laboratory Accreditation*, Chapman & Hall, London, 1995.

Data Analysis

Miller, J. C. and Miller, J. N., *Statistics for Analytical Chemistry*, 3rd Edn, Ellis Horwood, Chichester, UK, 1993.

SI Units and Physical Constants

SI Units

The SI system of units is generally used throughout this book. It should be noted, however, that according to current practice, there are some exceptions to this, for example wavenumber (cm^{-1}) and ionization energy (eV).

Base SI units and physical quantities

Quantity	Symbol	SI unit	Symbol
length	l	metre	m
mass	m	kilogram	kg
time	t	second	s
electric current	I	ampere	A
thermodynamic temperature	T	kelvin	K
amount of substance	n	mole	mol
luminous intensity	I_v	candela	cd

Prefixes used for SI units

Factor	Prefix	Symbol
10^{21}	zetta	Z
10^{18}	exa	E
10^{15}	peta	P
10^{12}	tera	T
10^{9}	giga	G
10^{6}	mega	M
10^{3}	kilo	k

(continued overleaf)

Prefixes used for SI units *(continued)*

Factor	Prefix	Symbol
10^2	hecto	h
10	deca	da
10^{-1}	deci	d
10^{-2}	centi	c
10^{-3}	milli	m
10^{-6}	micro	μ
10^{-9}	nano	n
10^{-12}	pico	p
10^{-15}	femto	f
10^{-18}	atto	a
10^{-21}	zepto	z

Derived SI units with special names and symbols

Physical quantity	SI unit		Expression in terms of base or derived SI units
	Name	Symbol	
frequency	hertz	Hz	$1\ \mathrm{Hz} = 1\ \mathrm{s}^{-1}$
force	newton	N	$1\ \mathrm{N} = 1\ \mathrm{kg\ m\ s}^{-2}$
pressure; stress	pascal	Pa	$1\ \mathrm{Pa} = 1\ \mathrm{N\ m}^{-2}$
energy; work; quantity of heat	joule	J	$1\ \mathrm{J} = 1\ \mathrm{N\ m}$
power	watt	W	$1\ \mathrm{W} = 1\ \mathrm{J\ s}^{-1}$
electric charge; quantity of electricity	coulomb	C	$1\ \mathrm{C} = 1\ \mathrm{A\ s}$
electric potential; potential difference; electromotive force; tension	volt	V	$1\ \mathrm{V} = 1\ \mathrm{J\ C}^{-1}$
electric capacitance	farad	F	$1\ F = 1\ \mathrm{C\ V}^{-1}$
electric resistance	ohm	Ω	$1\ \Omega = 1\ \mathrm{V\ A}^{-1}$
electric conductance	siemens	S	$1\ \mathrm{S} = 1\ \Omega^{-1}$
magnetic flux; flux of magnetic induction	weber	Wb	$1\ \mathrm{Wb} = 1\ \mathrm{V\ s}$
magnetic flux density; magnetic induction	tesla	T	$1\ \mathrm{T} = 1\ \mathrm{Wb\ m}^{-2}$
inductance	henry	H	$1\ \mathrm{H} = 1\ \mathrm{Wb\ A}^{-1}$
Celsius temperature	degree Celsius	°C	$1°\mathrm{C} = 1\ \mathrm{K}$
luminous flux	lumen	lm	$1\ \mathrm{lm} = 1\ \mathrm{cd\ sr}$
illuminance	lux	lx	$1\ \mathrm{lx} = 1\ \mathrm{lm\ m}^{-2}$

Derived SI units with special names and symbols *(continued)*

Physical quantity	SI unit		Expression in terms of base or derived SI units
	Name	Symbol	
activity (of a radionuclide)	becquerel	Bq	$1\ Bq = 1\ s^{-1}$
absorbed dose; specific energy	gray	Gy	$1\ Gy = 1\ J\ kg^{-1}$
dose equivalent	sievert	Sv	$1\ Sv = 1\ J\ kg^{-1}$
plane angle	radian	rad	1^a
solid angle	steradian	sr	1^a

[a] rad and sr may be included or omitted in expressions for the derived units.

Physical Constants

Recommended values of selected physical constants[a]

Constant	Symbol	Value
acceleration of free fall (acceleration due to gravity)	g_n	$9.806\ 65$ m s^{-2} [b]
atomic mass constant (unified atomic mass unit)	m_u	$1.660\ 540\ 2(10) \times 10^{-27}$ kg
Avogadro constant	L, N_A	$6.022\ 136\ 7(36) \times 10^{23}$ mol^{-1}
Boltzmann constant	k_B	$1.380\ 658(12) \times 10^{-23}$ J K^{-1}
electron specific charge (charge-to-mass ratio)	$-e/m_e$	$-1.758\ 819 \times 10^{11}$ C kg^{-1}
electron charge (elementary charge)	e	$1.602\ 177\ 33(49) \times 10^{-19}$ C
Faraday constant	F	$9.648\ 530\ 9(29) \times 10^4$ C mol^{-1}
ice-point temperature	T_{ice}	273.15 K [b]
molar gas constant	R	$8.314\ 510(70)$ J K^{-1} mol^{-1}
molar volume of ideal gas (at 273.15 K and 101 325 Pa)	V_m	$22.414\ 10(19) \times 10^{-3}$ m^3 mol^{-1}
Planck constant	h	$6.626\ 075\ 5(40) \times 10^{-34}$ J s
standard atmosphere	atm	$101\ 325$ Pa [b]
Stefan–Boltzmann constant[c]	σ	$5.670\ 51(19)$ W m^{-2} K^{-4}
speed of light in vacuum	c	$2.997\ 924\ 58 \times 10^8$ m s^{-1} [b]
Wien constant	b	2.898×10^{-3} m K

[a] Data are presented in their full precision, although often no more than the first four or five significant digits are used; figures in parentheses represent the standard deviation uncertainty in the least significant digits.
[b] Exactly defined values.
[c] Also known as the Stefan constant.

The Periodic Table

1 H																	2 He
3 Li	4 Be											5 B	6 C	7 N	8 O	9 F	10 Ne
11 Na	12 Mg											13 Al	14 Si	15 P	16 S	17 Cl	18 Ar
19 K	20 Ca	21 Sc	22 Ti	23 V	24 Cr	25 Mn	26 Fe	27 Co	28 Ni	29 Cu	30 Zn	31 Ga	32 Ge	33 As	34 Se	35 Br	36 Kr
37 Rb	38 Sr	39 Y	40 Zr	41 Nb	42 Mo	43 Tc	44 Ru	45 Rh	46 Pd	47 Ag	48 Cd	49 In	50 Sn	51 Sb	52 Te	53 I	54 Xe
55 Cs	56 Ba	57 La	72 Hf	73 Ta	74 W	75 Re	76 Os	77 Ir	78 Pt	79 Au	80 Hg	81 Tl	82 Pb	83 Bi	84 Po	85 At	86 Rn
87 Fr	88 Ra	89 Ac	104 Rf	105 Db	106 Sg	107 Bh	108 Hs	109 Mt	110 Uun	111 Uuu	112 Uub						

58 Ce	59 Pr	60 Nd	61 Pm	62 Sm	63 Eu	64 Gd	65 Tb	66 Dy	67 Ho	68 Er	69 Tm	70 Yb	71 Lu
90 Th	91 Pa	92 U	93 Np	94 Pu	95 Am	96 Cm	97 Bk	98 Cf	99 Es	100 Fm	101 Md	102 No	103 Lr

Index

Note. The use of an asterix * indicates that a definition/description of the entry is given in the text.

CPSIA information can be obtained at www.ICGtesting.com
Printed in the USA
LVOW05s0235050514

384440LV00001B/200/P